Commodities Pricing and the Bulk Trap

Fredrik Andrén-Sandberg

Commodities Pricing and the Bulk Trap

Learnings from Industries at the Forefront

Springer

Fredrik Andrén-Sandberg
Flyinge, Sweden

ISBN 978-3-319-72467-6 ISBN 978-3-319-72468-3 (eBook)
https://doi.org/10.1007/978-3-319-72468-3

Library of Congress Control Number: 2017962291

Printed on acid-free paper

This Springer imprint is published by Springer Nature
The registered company is Springer International Publishing AG
The registered company address is: Gewerbestrasse 11, 6330 Cham, Switzerland

Preface

Throughout the commodity market, major changes are visible on the horizon that might present both significant opportunities and threats for companies and countries present in the value chain—namely decommoditisation. The purpose of this book is to describe the trend based on experiences from the energy sector, which has already seen partial decommoditisation of their former commodities. From this case study, a theoretical framework will be established which can be applied across other sectors and utilised to create a better understanding on how best to respond to the changing commodity landscape—with the aim to reflect what is happening.

The reason for writing the book was due to scarce literature produced on the topic. Most literature available has primarily focused on either commodity trading or subparts of the decommoditisation trend such as the evolution of the of food sector, the renewable progression of the energy sector and the development of fair-trade goods. Up-to-date literature which sums up the trend across industries, while also drawing insight from those sectors which are furthest ahead, has, however, been rather scant.

One possible reason for the lack of written information on the subject may be due to the rapid changes in market conditions which the industries are currently undergoing. Hence, most of the knowledge available within the business community simply stays within their respective business community since time is too scarce for people caught up in the trend to sit down and reflect on the topic or for that matter writing a book on the subject.

The purpose of this book has thus been to alleviate the lack of publicly available literature and to cover the workings of this trend so far, as well as present consideration on how companies, countries and consumers could best take advantage of this new growth opportunity.

The target audience has primarily been business professionals and university scholars within business strategy, regulatory, marketing and supply chain management—as well as policymakers, politicians, NGOs and students active within similar fields. But it may also be of interest to the general population who likes to keep up with the latest economic developments within the field.

Thanks to my dear wife Maria and daughters Isabella and Matilda, who have given me the energy and allowed me the spare time to finish this book. A special thanks to my father Åke, a keen writer himself who has been supporting and patiently proofreading many versions of this book. Also a general thank you to all of my family, friends, colleagues and Linti—who has been supportive of this work.

Introduction

Commodities have played an important role in human civilisation for thousands of years and never before have commodities been so widespread in terms of volumes or people using them as they are today. Standardisation of products has in itself greatly increased the efficiency of production, while enabling commodity markets to increase the efficiency of trade with standardised goods being shipped freely to wherever they reap the highest economic benefit.

Commodity flows increased rapidly in the 2000 commodities boom in terms of both sheer volumes and origin diversity, which increased substantially as new sources previously economically inaccessible opened up to be traded across the globe.

There are some indications, however, that this trend has been hit by a counterweight, which may level out or even reverse the commoditisation of standardised products. The global penetration of the Internet in general and social media in particular has further empowered consumers to figure out where their commodities originate from and what externalities they cause, thus creating a differentiation in value of the commodities which hence cease to function as pure commodities and become products. There are reasons to believe that this trend will continue and even accelerate, with ever-increasing information density in products and services, and the prospect of the 'Internet of things', where most products and perhaps even commodities will be connected into a global information network.

Just as the introduction of commodities, the end of commodities might turn out to be a very productive evolution path for the global economy—where externalities are better priced into the products, enabling producers who minimise their negative external impact (and market their positive impacts) and reap higher margins for their produce. At the same time, producers who care less about negative externalities will not be able to do so, thus creating value for individual consumers while at the same time increasing overall value for the greater society by reducing the cost of producing those services and products.

This trend may prove to be a great opportunity to empower consumers, cause less harm to the planet as well as enable producers to fetch higher returns on their investments through products differentiation.

General Structure of This Book

The first chapters of the book will explain what commodities are and put them into a historical perspective to create an understanding on why commodities are commodities. This will create the necessary knowledge foundation to understand the main storyline of the book. From this baseline, the book then presents the main case study on how decommoditisation has progressed within the energy sector.

From the energy sector case study, a theoretical framework is presented and hypothesis formed implying that the underlying decommoditisation drivers should be present across other sectors as well. In the following chapters, I will examine the decommoditisation progression in other sectors through the lens of the framework and conclude what the implications for stakeholders might be and put forward suggestions on how to respond to these from a business and policy angle.

In the final chapters, some predictions on the trend including alternative paths are presented as well as a conclusion wrapping up the main takeaways of the book.

However, the experience does not need to end with the final words of this book. You are also welcome to join the discussion and give your own input on the topic at EndOfCommodities.com or at the Facebook group EndOfCommodities. Looking forward to meet you there!

Contents

Introduction to Commodities 1

To examine decommoditisation, we first need to understand the process of commoditisation and to pin down what attributes that actually defines commodities. In order to do this, we will dig into the history of commodities dating back to ancient times, and have a look at the golden age of commoditisation and examine the economic rationale for the commoditisation process. We will also briefly take a look of modern day commoditisation, to see what kind products are currently being commoditised, and examine what drivers are currently pushing this process.

1.1 How to Define Commodities

A commodity is a product, but what sets it apart from the conventional products category is that it is indistinguishable from any other product of the same type in the eyes of the purchaser, regardless of its origin. Raw materials elements such as copper, iron or gold are examples of this—since buyers are not able to distinguish a gold molecule mined in South Africa, from a gold molecule mined in North America at the time of purchase.

What also characterises a commodity is that the production volumes are sufficient to establish a market where the commodity can be traded. The price of the commodity then becomes dependent on the total production volume of that commodity on the market (together with the willingness to buy); interlinking single units of that commodity to every other unit of that commodity. Price is hence not set by the single product itself, but rather as a supply and demand curve for the entire commodity as such, essentially meaning that the price becomes a function of the total production volume and demand of the commodity.

Similar statements may be made for all products in some sense, but commodities still contrasts to normal products which can be differentiated with brands, inherent qualities, slight differences in design and other consumer determined characteristics. An example of normal products would be cell phones or suits; in

F. Andrén-Sandberg, *Commodities Pricing and the Bulk Trap*,
https://doi.org/10.1007/978-3-319-72468-3_1

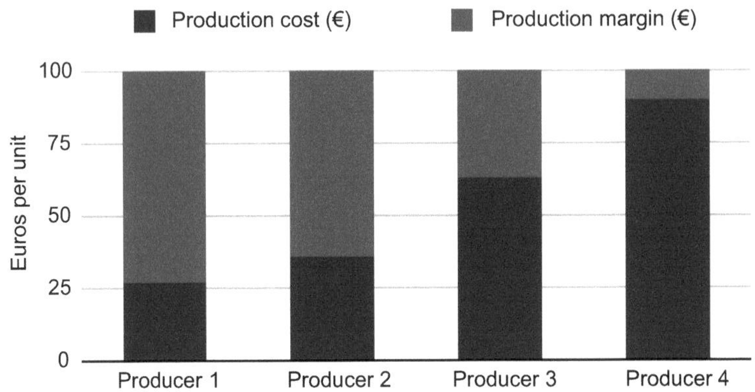

Fig. 1.1 Illustrative example of commodity logic. Showing market price for the imaginary commodity is set to EUR 100 per unit. Producer 1 will then receive a margin of EUR 70 per unit, while producer 4 will only reap a meagre EUR 15 per unit of that commodity. Highlighting the importance of efficient and low production cost in commodity markets since every producer fetch the same price for the produce

contrast to a ton of iron ore of a certain quality which probably would be priced independent of brand, design or who produced it.

All producers of the same commodity will thus at trading hubs be offered the same price for the commodity, independent of their production costs. Because of this indifference of production means in the eyes of the market, the profit a producer can turn on the commodity will be the delta between market price of the commodity and the production costs, including logistics, marketing, overheads, etc. This means that the producer with the cheapest cost of production will reap the highest economic benefit, as illustrated in Fig. 1.1. Therefore the market logic will force commodity producers to be as lean and efficient as technically possible, usually through economies of scale and high capital intensity, which brings down cost of the commodity and benefitting the overall economy.

Another definition of commodities is that they are the raw material building blocks that our civilisation tick; such as iron, coal, copper, wheat etc. This definition is true in the sense that most of the building blocks that are necessary for our civilisation to tick are traded as commodities. The reasons for this is that the sheer volumes of raw materials our civilisation demand have given rise to standardisation, efficient large scale production and commodities markets as a result. This definition is quite limiting though, since todays commodities is not only raw materials but increasingly also high tech products such as computer memory RAMs, and perhaps even internet bandwith in the future. Another objection to this definition is that all these commodities could actually be decommoditised and still be regarded as the building blocks of civilisation. However, while the latter definition of commodities as being the raw building materials that makes our civilisation tick might be limited. It is true that most products that form the basis of our economy is currently traded as commodities.

One could also define produce such as corn, copper, marble and rock in ancient times as commodities since they were also traded by the bulk for certain fixed prices. But as the volumes were much smaller and markets more local than today—they were probably experiencing higher degrees of differentiation. Tunisian corn may for instance have fetched different prices than Andalusian corn due to difference in consumer taste, and the price dependence between the two was probably not as directly dependent as they are today due to less developed logistical infrastructure.

The usual nomenclature of commodities used today is that 'soft commodities' in general are goods that are grown such as sugar, corn and livestock. Whereas 'hard commodities' are the ones mined or extracted such as iron, copper and gravel. Although the 'hard commodities' can be impacted by weather events such as heavy rain. In general though, 'the soft' commodities are the ones most prone to price fluctuations due to weather occurrences, such as drought or too little sun. Most hard and soft commodities are very price sensitive to global economic cycles due to their impact of marginal demand, which sets the commodity price.

On top of the hard and soft divergence, there is also the energy commodities such as oil, coal, power and natural gas; although some of these could fit into the hard commodity category as well, depending on what the classification aims at describing.

1.2 The Commoditisation Process and Its Economic Rationale

Commoditisation is closely connected with the process of standardisation, whereas custom production methods are replaced by standardised processes. This process have been the key to facilitate the industrial revolution and to encourage trade between countries and between companies.

One of the first breakthroughs in this process was made in the screw cutting industry in the late 1800s, where screws previously had been made in all kinds of thread sizes, angles, diameters and so forth. This meant that nuts and bolts from one company would almost certainly not be possible to be use with another company's nuts and bolts. However, by agreeing to standardise the dimensions of these products it opened up the path to interchangeability of production inputs and interchangeable parts, which in turn helps to facilitate the commoditisation of these products. The screws, nuts and bolts where not only standardised in terms of thread angles and diameters, but also in what the kind of metal used and its strength. This meant that a screw made from 'company A' was to the consumer indistinguishable from screws made from 'company B'. Furthermore, since the purchasers in theory are only interested in getting the lowest price possible for their purchase, the producing companies needed to focus on greater efficiency and to cut costs through modern technology and research in order to stay ahead of the game. This entailed very efficient production methods and large scale manufacturing industries, which meant large volumes of standardised products which could be traded in the market. These cheaper and commoditised products created large benefits for society as a whole, where tasks such as construction could be done cheaper, thus enabling economic growth through the build-out of large scale infrastructure projects and freed up capital and labour towards new sectors of the economy.

The standardisation also fostered increased trade and connections between companies and regions, causing interdependence between sectors of the economy—where products such as screws could now be manufactured by a 'specialised parts company' and sold to e.g. a train manufacturer who could be confident the products would meet their specifications. A well-known example of this trend is of course the modern production line made famous by the Ford company, where standardised parts and processes enabled the first mass produced car to be put in the hands of general consumers (Assembly lines was present well before Ford made them famous).

A second trend in this period of time was the introduction of standardised measurements of everything from weight and length to material strength and energy content of fuels—and the possibility to specify these data. This development was of large value for the commodity market, since the value of a commodity could now be measured in its actual value for the end utilisation. In terms of coal, the calorific heat content, water content and so forth is much more important than the actual tonnage or volume in the perspective of the end user. Although, the bulkiness and tonnage of materials were of course of large importance for the logistics of commodities, so these measurements continued to be important factors as well.

Before this period, it was very hard to agree on any measurement, and it was said that two men could not agree on the same scales even if their livelihoods depended on it.

It was still not easy though, as not all measurements where standardised into the convertible SI units—a heritage that still lives on today where we measure energy heating value in terms of: 'British thermal units', Megawatt hours, 'barrels of oil equivalents', joules, Kcal, and so forth. However, the important development was perhaps not the units themselves, but the standardised methods of getting the specification of the material and then convert it into any kind of interconvertible measurement. This enabled easier comparing of products and commodities, increased competition and facilitated more trade in between sectors of the economy.

1.3 The Golden Age of Commoditisation

Commoditisation can be defined as the process where products that are differentiated by quality, brand or some other differentiating factor ends up having the same value as similar products in the eyes of the purchaser, disregarding who produced them or where. The purchase prices for all products in the category becomes undifferentiated and manufacturers of the products loses their price setting power as the buyers will be seduced to buy only the cheapest products since considered generic, and the market moves towards perfect competition. The next step in this process is usually higher capital intensity and increased economies of scale due to intensifying competition, which furthermore lowers prices of the produced commodity. The intellectual capital required to make the commodity is usually diffused across the market over time, where capital sacristy and local conditions along with demand will be determining what margins producers can reap. The process also has a tendency to increase trade since it becomes more profitable and sensible to produce the goods in larger volumes where production costs are cheap, subtracting the cost of bringing the goods to market. Significant volumes must therefore be produced in order for a producer to stay competitive and commodities are thus usually shipped around the globe in bulk carrier shipping routes.

Few or no products could be considered perfect commodities, since there's usually some kind of differentiation. However, when market treats goods as equivalents or nearly so with no regard of who produced them, then it is usually considered a general commodity.

> *From the taste of wheat it is not possible to tell who produced it; a Russian serf, a French peasant or an English capitalist* —Karl Marx, commenting on commodities

The commoditisation process has occurred for several product categories, and major commodities today such as iron, copper and oil are good examples of 'commodities' where the price is very much determined as a function of supply and demand of the market as a whole. To compare this with general goods, the demand and price for a certain car model is arguably not directly dependent on the price and demand for other cars models. This can be disputed in terms of relative price setting, but in general there is a price disparity of what people are willing to pay for a BMW compared to a Toyota because in the eyes of the consumer, they are not the same.

The distinction between commodity and product should neither be seen as a binary distinction, and very few products are either fully fledged commodities or products. It is also the case that different consumers experience commodities different, where some may view food items such as milk and eggs as perfect commodities, where other consumers may see huge difference in terms of how the product tastes or even been produced.

Another trend that has boosted the development of commodity markets is the standardisation of shipping, the subsequent decreased costs of freight and increased trade due to removal of trade barriers, which is illustrated in Fig. 1.2. When the cost of moving goods from one point to another is high, the locally produced product or commodity increases its competitive advantage over goods produced in other parts of the globe. On the other hand, when the cost of moving goods is decreased, the local producer have to compete with similar products from far away on similar turns. An example of this is for instance sheep meat from New Zealand competing directly with local European produce. This increases the market size of the products, drives the rationale of efficiency of scale and subsequently has a boosting effect to the development of commodity markets.

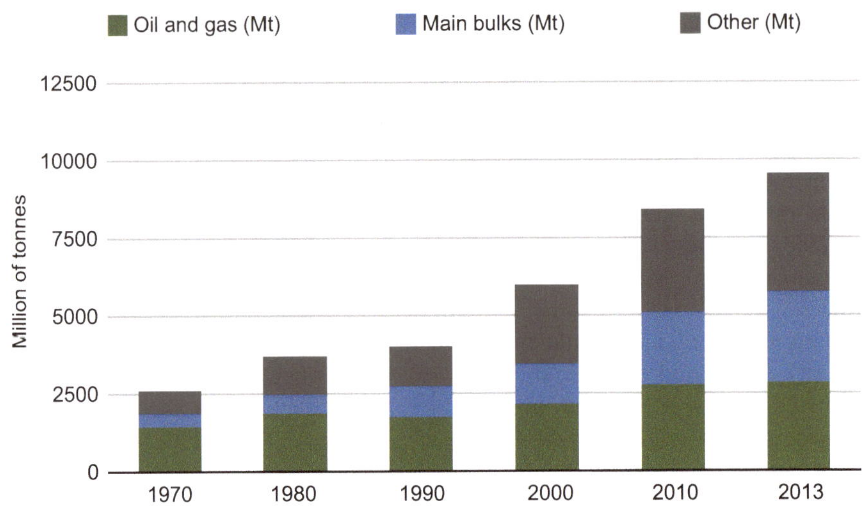

Fig. 1.2 International seaborne trade in million tonnes loaded. That has increased rapidly during the last 40 years, increasing the size of product markets and driving commoditisation. The lowering of import tariffs, the establishment of free trade agreements and the decreased cost of shipping have had a dramatic effect on shipped commodity volumes. Source: UNCTAD (2016), Review of Marine transport

The commoditisation trend has fostered larger ships that can ship even heavier bulk volumes which drives down the unit cost significantly of the goods being shipped. The largest bulk carriers today shipping oil are just short of 400 m long, 60 m wide and can ship 440,000 tons of crude oil per voyage. This can be put into context to the typical oil tankers of the World War II era, which were just 160 m long and shipped a maximum of 16,000 tons of oil. The evolution is shown in Table 1.1, with a 20-fold increase in ship capacity over the last 40 years. This significant increase in magnitude has thus reduced shipping costs, and the cost of shipping cargo has dropped by approximately 90% from 1956 to 2016. There has also been a standardisation of the non-bulk shipping units to in the form of the shipping container—enabling much faster loading and unloading at ports, but also easier transport on trucks to the end destination. This type of shipping has increased very rapidly, with the capacity of the global container ship fleet increasing by almost 30% just between 2012 and 2015. All these trends has dramatically lowered the cost of shipping, which has increased trade volumes and supported ever larger commodities markets.

1.4 Modern Day Commoditisation of Products

The question on where to draw the line between commodities and products can be, as already mentioned, a bit vague. However, if defining commodities as hardly differentiated products traded by the bulk of a given quantity and sold on a market

Table 1.1 Evolution of container shipping vessels. Source: Rodrigue et al. (2017), The geography of transport systems

Size classes/Years	Length (m)	Draft (m)	Capacity (TUE)
1956–1970	135–200	9	500–800
1970–1980	215	10	1000–2500
1980–1988 (Panamax)	250–290	11–12	3000–4000
1988–2000 (Post-Panamax)	275–306	11–13	4000–5000
2000–2005 (Post-Panamax Plus)	335	13–14	5000–8000
2006–(New Panamax)	397	15.5	11,000–14,500

where the producers has lost their price setting power; then there have been several partial commoditisation of products in the last decades. In the following examples there may still be some brand factor differentiator, but there is evidence that this is fading when the products are on their way of becoming pure commodities.

Memory RAMs for computers have a recent years become more and more commodity like, with end user seldom care which manufacturer has produced their 8 GB RAM. The cheaper the better, while quality and longevity of the products is just expected to be similar across the board. The commoditisation of computer memory has led to substantial price decreases per storage volumes, with price decreases of 100,000 times in the last three decades per storage volume basis as shown in Fig. 1.3. A driving factor of this commoditisation besides the large volumes is that memory RAMs does not have any moving parts, and therefore has less tendency to break down compared to products such as engines or kitchen appliances. This entails that quality and thus manufacturer is less of a factor for the consumer regarding RAMs than for instance a home backup diesel generator.

Photovoltaic solar panels is another product that is being ever more sold in bulk, where even spot market for silicon modules has been set up in several places. As with memory RAMs, silicon solar panels do not contain any moving parts and standardisation of the products has led to the main differentiating factor being effect per unit area and appearance. Besides these differentiators they are more or less sold on a cost per panel or per effect (kWp). Due to the commoditisation and scale of the industry, prices have fallen dramatically over the last decades as illustrated in Fig. 1.4. The price decreases has even been self-reenforcing; where increased production of panels has led to lower prices, and lower prices has led to increased demand and thus production volumes due to improving economics from the lower prices. This mechanism have had major impact on energy markets across the globe in the last decade which we will look into in the following chapters.

However, since manufacturers do have different specifications and efficiency on their panels, and that there is a burgeoning development of non-silicon solar panels such as thin film technology with different properties, there is still arguably some room for differentiation of the product. It is thus questionable to view PV solar panels as a complete commodity even though it is moving in the commoditisation direction.

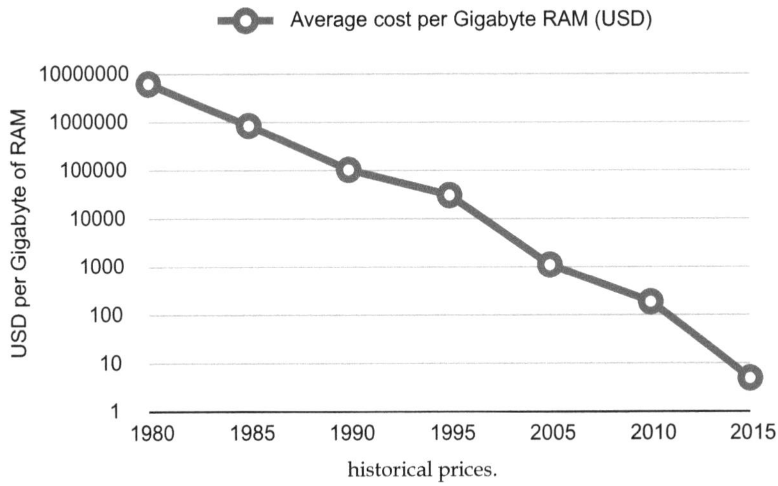

Fig. 1.3 Cost development for computer memory RAM. Source: Statistic brain (2017), Memory RAM historical prices

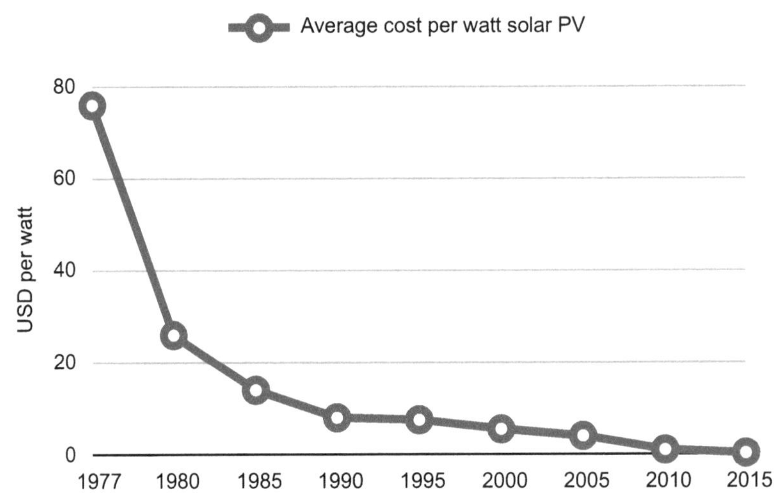

Fig. 1.4 Cost development for PV solar panels. Which has decreased 200-fold in just under 40 years. Source: Bloomberg New Energy Finance (2016), Evolution of PV solar panel cost

Consumables such as *multivitamins* could be considered commoditised, where a tablet of 100 mg vitamin C has the same value to most consumers disregarding who produced them. This will of course always be within the boundaries that the consumers trust the producer enough to such an extent that they are willing to eat it. An example where there has been a decommoditisation in this regard is within the baby formula market, where stories about tainted products with forbidden chemicals in China caused Chinese consumer to abandon native suppliers in favour to foreign ones.

The last example of current product commoditisation is that of *batteries*, which has become ever more important in our everyday lives in anything from computers to cellphones and recently also in cars. Lithium ion batteries which is the main technology used due to its technical characteristics such as high energy density to weight ratio, has seen substantial price reductions partly due to the rapid increase in electric vehicle usage, and even more recently energy storage systems. Batteries has thus from a utilisation perspective become ever more rated in terms of Euro per kWh, where the supplier of the battery or production methods does not seem to matter so much for the end consumers anymore.

These examples are just a few of the current products that are arguably going through a commoditisation process, but there are many more products which are currently losing their differentiation possibilities.

We now know what commodities are, and have also gained some overall knowledge on the commoditisation rationale and its progress throughout history. In Chap. 2 we will examine how the recent history during the pinnacle of commoditisation history—namely the commodity boom of the early twenty-first century has impacted the commodity market, perhaps even irreversibly going forward.

References

Bloomberg New Energy Finance (2016) Evolution of PV solar panel cost. http://www.bbhub.io/bnef/sites/4/2016/04/BNEF-Summit-Keynote-2016.pdf. Accessed 10 Sept 2017

Rodrigue J, Comtois C, Slack B (2017) The geography of transport systems. Taylor & Francis, Abingdon

Statistic brain (2017) Memory ram historical prices. http://www.statisticbrain.com/average-historic-price-of-ram. Accessed 10 Sept 2017

UNCTAD UN Conference on Trade And Development (2016) Review of marine transport. http://unctad.org/en/PublicationsLibrary/rmt2016_en.pdf. Accessed 10 Sept 2017

2 The Unlikely Seeds of Commoditisation Reversal

What started out as a very good era for commodities in the early twenty-first century, where both prices and volumes skyrocketed driven by a multitude of factors, may counterintuitively turn out to have sowed the accelerating seeds to its own dismantling. In this chapter we will gain some knowledge on what happened in the 2000s commodities boom, and try to understand what irreversible impact it may have had on commodity markets going forward. We will also look into just how considerable commodities and their prices actually affects global politics as well as the economic wellbeing of nations.

2.1 The 2000s Commodity Boom (2000–2014): Increased Differentiation of Commodity Sourcing

Current traded commodities plays a significant role in the overall global economy due to their large quantities and basic function in our societies. This can be seen through a time perspective where low prices for energy has been correlated with increased economic activity. A good recent example of the importance of commodity production and trade is the 2000s commodity boom, which had a significant impact on the global economy and societies.

In the early twenty-first century the global economy entered what has since been called the 2000s commodities boom, or the commodities super cycle. What happened was a dramatic increase in commodity prices, that led to an significant expansion of production as well as more diverse sourcing of commodities.

Global economic growth averaged almost 4% during these years (at this rate doubling global GDP every 18th year), with the newly coined BRIC economies (Brazil, Russia, India, China) with particularly China growing at an incredible momentum. This growth turned China from a backwater economy into the largest consumer of many major commodities by importing iron ore, coal and copper to fuel its booming market which recorded double digit growth for almost an entire decade. There was also in increased demand from developed nations for

F. Andrén-Sandberg, *Commodities Pricing and the Bulk Trap*,
https://doi.org/10.1007/978-3-319-72468-3_2

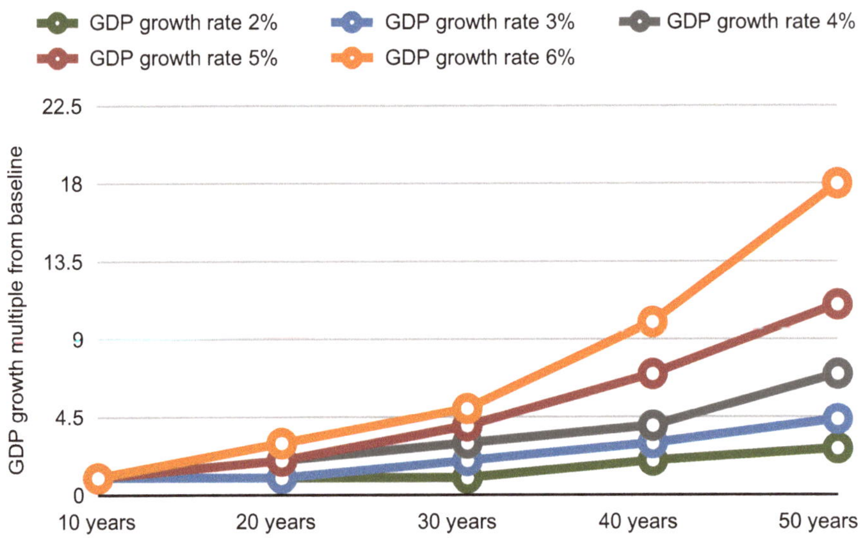

Fig. 2.1 Economic GDP output doubling at different growth rates. Where a 6% growth rate implies a whopping 18-fold increase in economic output over 50 years, while a 2% growth rate implies only a mare doubling of economic output

electronics, cars, and consumer products, further feeding the demand for commodities. An illustration how profoundly different growth rates translates into GDP levels can be seen in Fig. 2.1—where 6% growth rate implies almost seven times higher GDP levels compared to 2% growth rates over a 50 year period. Absolute economic growth rate correlate with commodity consumption growth, but not in a 1:1 relationship, and many advanced economies have been able to decouple economic growth from for instance energy consumption. In theory economic growth could continue indefinitely with ever smarter and circular resource handling systems, without increasing primary commodity consumption—but this has not been the case so far.

Due to the increased demand, prices on everything from chlorine and copper to grain and oil soared to new heights, with commodities such as the sulphuric acid (used as industrial component) increased by more than 300% within a year. Steadily increasing population growth also fuelled demand for grains and animal production in conjunction with increased demand stemming from the expansion of biofuels. Food production at this point in time was at an historical all time high, but the changed utilisation still caused supply and demand imbalance. Part from the underlying fundamental trends, there was short term reasons for the price hikes as well, such as bad harvests and droughts in several important countries including major grain producers such as Russia. Economic speculation may also have contributed to the sharp increase in basic food crop prices. Regardless the cause, the result was that the price of rice doubled in just under 7 months. Wheat prices also more than doubled within a year, and food riots broke out in several countries

Fig. 2.2 Oil prices USD per barrel. Showing large price increases and volatility during the early twenty-first century. Source: U.S. Energy Information Administration (2017a, b), West Texas Intermediate spot prices peaks and averages

including Egypt and Morocco, due to fear of not being able to afford the staple commodity.

Another particularly important trend during this era was the substantial increase in basic energy prices—where coal, natural gas, oil and electricity (all partially interlinked) experienced a drastic increase in prices. Spot oil prices went from a low of 20 USD a barrel in the beginning for 2000, to over USD 150 for a brief period of time in 2008, settling at around USD 100 in 2014—a whopping 500%–750% price increase of a commodity that played a fundamental role in the global economy. This trend can be seen in Fig. 2.2.

Interlinked to the main energy commodities, there was another energy source which experienced a spectacular price increase—namely uranium. This commodity saw a whopping 30-fold increase in its price from 2001 to its 2007 peak, substantially benefitting the main producers of the commodity. Similar trends could be seen for a large variety of commodities during this decade, partly due to the interlinkage for many commodities through various industries and usages. The geopolitical implications of the trend should not to be underestimated, where Russia, Gulf nations and other countries with large oil exports experienced an economic boom and increased global political clout which we will examine further later in this chapter.

What also happened during this period was that the increased prices brought forward an interest from developers and miners to tap into new sources of energy. 'Tar sand oil' in Canada had previously been uneconomical to extract from the ground due to much higher production costs than conventional oil wells, a fundamental that changed with the increased price of oil during this period. Similar economic hurdles had been true for American shale gas and oil as well—but the higher prices of energy brought forward nothing less than a revolution in these fringe fossil fuel markets.

2.2 The Collapse of the Commodity Bull Market

Part from a brief slump due to the 2008–2009 debt crisis, commodity prices peaked in 2011. Several reasons can be attributed to the commodities boom, such as demographic factors, strong economic growth, dwindling supplies, or the sevenfold increase in investors participation who's trade brought increased volatility to the commodities market. The post-financial crisis commodity price bubble has been particularly blamed for the latter. In regards to the cause for the collapse of the bubble, the opinions differ between scholars in the subject. Some economists have blamed the commodity bubble itself for its own collapse. This since the global economy simply could not sustain such a quick price elevation of primary commodities without being forced into a recession, which then in turn collapses the commodity prices down to more sustainable levels due to reduced demand. There are some merit to this theory, as low commodity prices historically have had a positive impact on economic growth since less resources has to be spent on basic inputs, and more can be diverted to down stream value adding activities. It is probable that this played part in the collapse, but also likely that it was a combination of different factors.

Regardless the cause of the bubble bursting, there were plenty of investment firms, funds and individuals who suffered substantial losses by investing in commodities prior to the boom, just to see their investments eroded as prices collapsed. Another obvious looser of the collapse was the industry extracting or producing commodities, which had anticipated elevated prices levels and demand for years to come, and had thus pursued investment decisions according to those business assumptions which turned out to be flawed.

Many of the producers thus sat with unprofitable investments which could not cover their capital- or even operating costs, while being stuck with an industry overcapacity that continued to have a negative impact on commodity prices for years to come. Mines had to close across the globe, both in developed nations such as Sweden and Canada, and in more commodity exposed countries such as Zambia and the Democratic Republic of Congo in order for the market to rebalance. This incurred substantial impairments for companies, and economic suffering for individuals affected.

Disregarding the reason for the boom and the effects of the collapse, one of the main consequences of the price boom was an increased demand for commodities. This in turn motivated financiers to pour investments into new sources of production to supply the increasing demand in the market. This rush for resources allowed for several unconventional sources of commodities to emerge which had not been economically rational to bring online before the commodity price increase, due to higher production costs compared to the conventional sources. This source diversifications had profound ramifications which we will investigate further in the following sections and chapters, especially since it entailed new externalities that became increasingly visible and vividly debated in the public domain.

2.3 Lasting Consequences of the Commodity Boom: New Sources Coming Online

One of the new sources of fossil energy coming online was US shale oil and gas brought forward by hydraulic fracturing of rock or for short 'fracking'—which was made economical viable by the increased prices for natural gas and oil. In short, it works by pressing water and chemicals into a drilled hole in the ground, where the increased pressure make the ground rock fracture and thereby release gas and oil which is stored within it in geological formations. The environmental consequences of this practice have been highly debated, but what has been less debated is the substantially higher cost of retrieving this fossil energy compared to many conventional oil and natural gas fields. Due to the local nature of the practice, the wells are relatively small and run out of their resources quickly, meaning that new wells must be drilled continuously in order to maintain production volumes. This generally entails increased CAPEX and OPEX costs per oil volumed produced compared to conventional oil wells, and puts the cost of producing oil and gas from shale formation in the range of USD 30–70 per barrel of oil or even higher. For context, a Saudi barrel of oil has a production cost just around USD 10 per barrel of oil.

Due to prices hovering just around USD 100 per barrel of oil and the 'ease' of doing these kind of business in the United States, the production of shale oil and gas exploded. This development had a large impact on the global fossil fuel landscape as well as the US domestic one, where oil imports could be reduced on behalf of the domestically consumed commodity—which can be seen in Fig. 2.3. Part from oil, the shale revolution also expanded the US natural gas supply which reduced prices substantially. This altered the logic of the US power market which traditionally had been very reliant on coal, to tilt towards new gas-fired generation instead. Unintendedly this had a very positive effect on US greenhouse gas emissions since coal-fired power stations which emits more carbon dioxide per electricity produced than gas-fired alternatives where being closed down in favour of the latter.

Another new source of oil coming online due to the high commodity prices in North America was the Canadian tar sands. Tar sand is actually just oil that has migrated to the surface from below in a natural process, and which has then mixed with the soil in the ground. Due to the proximity to the surface, many of the lighter elements of the oil has evaporated and was is left is a rather thick blend of oil mixed with other minerals and sand.

In order to extract these oil fractions, large areas of land is turned into something that looks like shallow open pit mines, causing severe environmental impact where extracted. Since a lot of effort is also needed to extract these fractions, plenty of energy and investments are spent in the process, which means that the energy input-to-output is a lot lower than for traditional oil fields. Most tar sands needs roughly

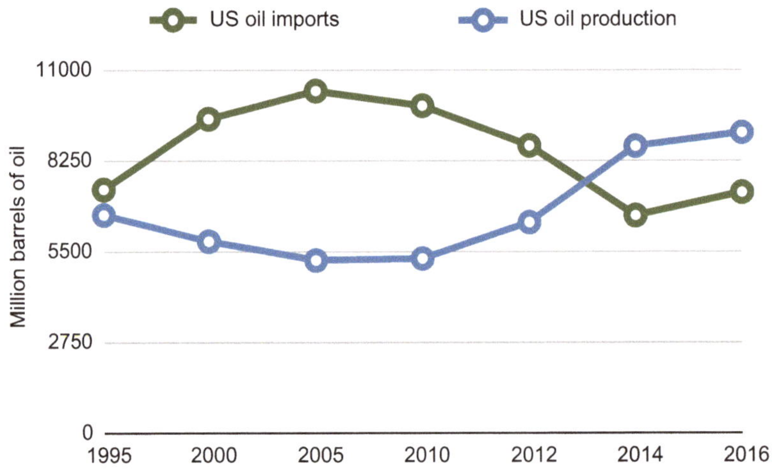

Fig. 2.3 US oil production and imports. The shale gas revolution have produced significant home grown oil volumes in the USA, and have had a drastic impact on US net oil imports and domestic production volumes. Source: U.S. Energy Information Administration (2017a, b), Crude oil production and trade

1/5 of the energy extracted for its own use to be able to recover it. This also mean that the well-to-wheel CO_2 emissions per barrel of oil is considerably higher than for conventional oil. In terms of economics, the cost for production varies considerably, but some projects have been estimated to require oil prices to stay above USD 80 per barrel of oil to remain profitable, although other projects reports profitability of just above USD 50 per barrel of oil.

A third oil fraction that also gained increased attention in terms of exploration were oil fractions which are harboured in hard to reach sea regions such as the Arctics. Just as the previous examples, the off-take price per barrel of oil need to be very high in order to investments in these fringe oil-fields to pay off. The price differs from project to project, but estimations for Arctic oil production cost amounts to something in the order of USD 100 per barrel of oil. With the downside of severe risk to local ecosystems, and limited possibilities for cleanup operations if a disaster such as an oil spill would strike.

The trend of new commodity sources of different 'qualities' and externalities has, however, not been limited only to oil or gas production, but has also taken emerged in the extraction of coal and minerals for example in Australia. It also increased profitability for agricultural and farm products, which enabled increase in land use for sugar and soy bean plantations in countries such as Brazil, and palm oil plantations in countries such as Indonesia. Both examples with arguably high externality costs due to displacement of local ecosystem and fauna, as rainforests are being cleared to make way for large scale plantations and agricultural land.

A somewhat positive impact of the commodity boom has been in terms of world income distribution, where small scale farmers in the developing world and producers of other primary resources have been able to increase their earnings and better invest in their own future. A lot of commodity resources was also located in countries much economically dependent on that single commodity, which made for much needed boost for those national economies.

The flip-side is that extraction of precious metals and minerals in poor and conflict ridden areas such as eastern Congo has also become more profitable, and thus able to further fuel conflicts already present.

2.4 The Geopolitical Impact of the Bust Bubble

The large inputs of new commodity sources in conjunction with global economic slowdown and geopolitical interventions did eventually cause prices to drop however. Russia and the oil producing Gulf countries had during the commodity boom experienced a significant economic upswing since their economies was in large part being driven by fossil fuel exports. However, oil prices dipped significantly in the mid 2010s after Saudi-Arabia refused to cut production in order to support prices as it usually had in the past, responding to sliding oil production market share in favour of American shale oil producers. This triggered an oversupply of oil in the market, lowering prices and forcing American shale gas producers out of business in the medium turn, thus helping Saudi-Arabia to regain its lost market share.

However, the flip-side of the strategy was that if prices where to once again increase, US shale oil producers could just increase production and stabilise prices due to very quick shale well deployment times—thus setting a de-facto global price ceiling on oil based on US shale oil production costs instead of a price floor.

Nevertheless, the ramifications of the oil price fall for several producing nations was significant, where drastically diminishing oil incomes in Russia has had geopolitical consequences after putting pressure on the state finances. In other oil producing countries such as Venezuela, Nigeria and Iran-oil prices were less than half of the prices needed to balance and sustain the governments budgets. This caused economic turmoil and even deep recessions in countries such as Brazil and Venezuela. In the latter, electric power in 2016 needed to be rationed even though the country possess the largest oil reserve in the world. Even Saudi Arabia, the kingmaker in the global oil market and partly responsible for the reduction in prices, began to suffer significantly from the lower oil prices. As their budget recorded an approximately 17% deficit in 2016—compared to a ~30% budget surplus in 2008, significantly limiting the manoeuvring room for the rich country.

The winner of low commodity prices was arguably regions such as India and the European Union, who were large net importers of both energy and other raw material commodities. The European Union for instance imports 53% of all energy consumed, costing more than 1 EURbn per day or 20% of total imports costs. Imported share amounts to 90% of its crude oil consumption, 65% of natural gas consumption, 40% of coal and solid fuel consumption and 40% of nuclear fuels. Approximately half of EUs consumption of gas and 40% of its oil stems from Russia and Norway. The fluctuating import cost for Oil in particular for the EU can be seen in Fig. 2.4.

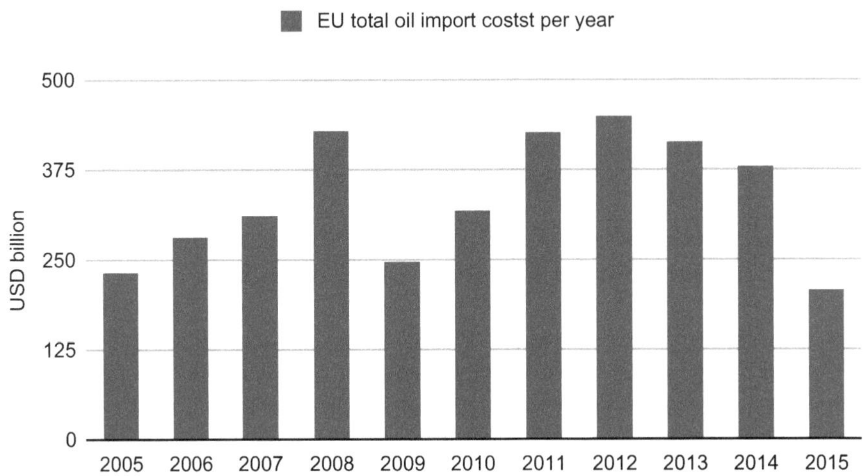

Fig. 2.4 EU total oil import costs. Source: European commission & Eurostat (2017), EU Crude Oil Imports and supply

Net commodity importers such as the EU and India did not benefit much from high commodity prices, since taking a significant toll on their import bills. Due to the lower commodity import costs, the regions were able to use the surplus cash for other investments or just general consumption instead.

However, for the EU the lower commodity prices did imply a lower import bill, but since many of the large companies within global commodity extraction such as Shell, Rio Tinto and BP are based in the EU—the lower commodity prices was somewhat of a mixed blessing. Since also coinciding with a rather deep recession, the lacklustre commodity prices was also a cause of concern since it put a downward pressure on inflation, and could have led the region into a deflation cycle.

In the case of India, gasoline had also been subsidised which meant a big relief on public finances once the high commodity prices was a thing of the past, also freeing up resources to invest in more domestic growth instead. The lower prices also offered relief for net commodity importing countries such as Egypt. However, in the case of Egypt the lower oil price also implied lower remittances from Egyptians working abroad in the oil industry—so the net logic of who benefits from commodity price movement is seldom completely as simple as one would first imagine.

The 2000 commodity boom had a substantial impact on commodity markets, in particular for commodity production where the spectrum of sources was significantly broadened. This development had both positive and negative consequences for the regional economies and their respective natural and social environment. However, the increased volumes and broadened spectra in commodity production in conjunction with information technology trends, might be the catalyst which causes a reversal in the commoditisation trend as we will further investigate in Chap. 3.

References

European commission & Eurostat (2017) EU crude oil imports and supply cost. http://ec.europa.eu/energy/en/data-analysis/eu-crude-oil-imports. Accessed 10 Sept 2017
U.S. Energy Information Administration (2017a) Petroleum and other liquid spot prices. https://www.eia.gov/dnav/pet/pet_pri_spt_s1_a.htm. Accessed 10 Sept 2017
U.S. Energy Information Administration (2017b) Crude oil production and trade. https://www.eia.gov/petroleum/data.php. Accessed 10 Sept 2017

3 Break in the Commoditisation Trend: Information Proliferation Starting to Shatter the Commodity Illusion

Technology has undoubtedly had a pivotal impact on production of goods and commodities throughout history. From shipping's ability to create large interconnected markets, to excavation technologies churning out impressive amounts of raw material for those markets. Technology has arguably been the primary enabler of commodities themselves, and been pushing the trend for increased commoditisation of products through standardisation and ever rising volumes. However, the coming impact of technology on commodities might entail somewhat different consequences. Whereas the historical commodity impact from technology mainly have been in form of ever more sophisticated and larger scale machinery, the coming technology trend will be more focused on information creation and proliferation. Since information proliferation may turn out to be perfect counterweight against commoditisation, the new technology revolution might have an irreversible impact on commodity markets going forward. Which we will examine in this chapter.

3.1 What Could Possibly Counteract the Commoditisation Trend Logic

The inherent problem with commodities is that the indistinguishability of the goods is largely an illusion, conjured since the end customers are unable to fully comprehend all parts of the goods inherent traits. If the customers could see next to the store window how the diamond for sale has been mined, or in which manner the meat at the butcher shop was raised, then there would arguably be a differentiation in prices for the supposed commodity. The same would arguably also be true for other commodities such as iron and oil, where different modes of production diverge substantially in terms of environmental impact-as it's a bit messier to produce oil from tar sand in Canada than it is to extract oil from a Kuwaiti oil field.

These differences is generally reflected in the production costs of the commodity, but is completely concealed to the end consumer purchasing the goods, which

F. Andrén-Sandberg, *Commodities Pricing and the Bulk Trap*,
https://doi.org/10.1007/978-3-319-72468-3_3

hence fetches the same end consumer price, albeit at different margins for the producers.

This imperfect sharing of information has largely been possible due to the detachment of production from consumption in the eyes of the consumer, and lack of information flow and transparency in the commodities market. Most customers seldom consider that the contents of carton of milk is actually originating from the udders of a bunch of cows. They probably rarely even contemplate what kind of factory the container comes from or what kind of conditions that were present there. Rather, in the eyes of the consumers—milk comes from the store. It has a certain brand, sometimes being represented by a monkey instead of a cow, and it may taste like strawberry or chocolate depending on which colour of the container you choose.

This is a very recent notion; just a 100 years ago in the developed world and still in many parts of the developing world, people knew exactly where their milk came from—since its origin was from their own cow in their yard or the farmers next door. As they had this information, they would also know more about the production methods of that milk, and might have had an opinion on how the cow should be treated while providing the resource. Arguably, many consumers would probably still care if they just had access to the same kind of information today. But since they do not, milk is mostly considered a semi-commodity where the specifications of fat content, flavour and price is the defining factors whether a purchase will go through at the store or not.

However, if the only thing keeping commodities to function as commodities is the 'concealment' of information from the consumer (not implied as a conspiracy, just a market state)—then the main thing that would decommoditise commodities will have to be the proliferation of that information.

This lack of knowledge regarding commodity externalities in the market is seemingly about to change with the trend of ever increasing information density in all products. A trend which is bound to have substantial impact when reaching a pivot in the commodities market.

3.2 Increased Public Awareness of Production and Supply Chain Externalities

Would someone in the USA be willing to pay more for US oil than for Saudi oil? The answer is arguably 'yes' for some consumers, be it due to patriotism, Saudi regime criticism or some other personal view that the end customer might possess.

The oil commodity would hence be valued differently by the end consumer if they would receive the right kind of information, and would perhaps even be willing to pay a premium price due to that information for a certain version of that commodity. Today it is usually not possible for consumers to make this choice of commodity supplier in the event of gasoline purchase, but there is already an increased awareness and differentiation in commodity prices ongoing in everything from energy and food, to diamonds and clothing.

Commodities versus products has never been a binary definition, and there has always been a spectrum of how commoditised a product is and to whom. Solar energy may be worth more than conventional energy to some, but not at all to others. Food items such as eggs from free roaming hens bring additional value for some consumers, while not differentiated at all to some industrial bakeries if they cannot transfer some of the value to their end customers. Organic products fetch higher value for some consumers, but not for others, etc. This spectrum of consumer preference has always been present in the market, what has arguably changed now is perhaps not public sentiment as such, but rather the ease of which information of the product or commodity could be gathered and thus the customer ability to make a 'better' informed decision on what to purchase.

Technology and the proliferation of information brought in by classical media such as television, magazines and books have played a large part in this trend—where documentaries and news about whaling, blood diamonds or destructive fishing practices has raised customer awareness to a point where producers and distributors of the commodities have been forced to take action.

Trends in public sentiment can take hold quickly, and was partially demonstrated when France test detonated nuclear devices in the 1990s, where French wine export plummeted due to global consumers using their purchasing power to punish the country for its environmental violations. France later stopped the test detonations and has worked against nuclear testing above ground since. Wine can certainly not be seen as a commodity, but it still shows the power of consumer purchasing behaviour when taking collective action against perceived unfair production methods, or in this case a perceived immoral policy of the producing country. The questions is if other French commodities exports would also have been hit, if the consumers in other countries could choose to not buy French petroleum products, steel or some other French commodity export.

Food production has also come under increasing scrutiny due to animal welfare practices, water usage, land degradation and the impact on local communities and environment from forest clearing. The sustainability of fishing practices are being questioned as sea animal population has plummeted, and even the origins of diamonds have begun to cause distress for some consumers.

Another externality concern that has spread rapidly and broadly is the one caused by the energy industry. This since it spews out billions of tonnes of carbon dioxide yearly into the atmosphere, causes local air pollution such as smog and acid rain as well as causing substantial impact to the land while extracting fossil fuels. Other consumers are concerned about the nuclear waste being produced by nuclear power stations, others about the landscape degradation from wind turbines polluting the landscape picture. This concern have caused the industry to change irreversibly as we will examine in the following chapters, but the main takeaway is that there is a general overarching trend of ever increasing spread of externality information, which is forcing markets to transform.

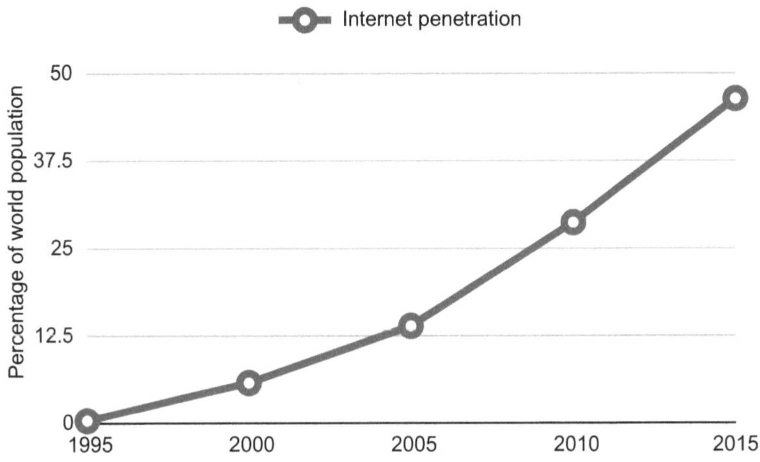

Fig. 3.1 Global internet penetration percentage. With rates growing towards saturation, which has significantly changed the information proliferation landscape. Source: United Nations Data (2017), World development internet users

3.3 Social Media and Digitalisation Impact on the Differentiated Commodity Market

The trend that modern media started has been reinforced, and arguably even revolutionised by the latest trends of digitalisation, internet and the rise of social media as a form of communication. The dramatic growth in global internet penetration rate can be seen in Fig. 3.1. Awareness of production practices can now be spread quickly peer-to-peer through social media, pictures though Instagram, news articles virally shared though Facebook, and cell-phone captured videos distributed on youtube.

Consumers have now become empowered to use the information they gather to arise awareness among other consumers by distributing it through their networks quickly and with little effort. Problematic issues of production practices can easily be shared, just as the alternatives to those production methods can be highlighted. Disturbing working conditions of textile workers used by a clothing retailer, can in the same social media post be paired with an alternative retailer showing off their positive credentials (with textile workers enjoying their weekly lunch massage). Hence, supplying the consumer with a solution to continue purchasing the product they demand but with the benefit of a better conscience, although possibly attached to higher price tag.

As with every technology, the trend also comes with it's own risks—such as disinformation which can also be shared and proliferate with never before seen efficiency. This trend was very vividly debated in during 2016 as disinformation and so called fake news started to impact democratic processes in several countries. Equally, disinformation and slandering regarding commodities and products could easily be shared viral though social networks without the control by accredited people. This problem has yet come to a solution, but some of the main internet operators such as Google and Facebook has as of 2017 started to tackle the issue though several initiatives.

Even with the obvious problems and current flaws facing the information sharing technology, there are no signs of the peer-to-peer trend of sharing of information is abating in any way. Rather it seems to be gathering momentum, and already have a large impact on marketing activities of brands and producers of goods. It is therefore good arguments on why that the trend should be taking increasing hold in the commodity sectors as well and there are several examples of this that will be further investigated in the coming chapters.

3.4 Companies Push to Escape the Bulk Trap

Most companies fear, and are doing what they can to avoid so called "bulk trap". This situation is where the goods being produced is the same as anyone else that can produce the same specifications, from the view of the market. It thus becomes very difficult for companies to differentiate their products, and the main action they can undertake to improve their situation is just to increase their volumes (if having positive margins) or lower their costs base if technology allows. However, these actions should theoretically just lower the price of the product or commodity throughout the market, which will force other companies to also increase volumes or reduce production costs and thus depress the price even further. Even worse, if companies find themselves incapable of keeping up with their peers, they will be forced to close down their production due to the stiff competition in order for the market to rebalance to the new equilibrium.

You don't have to be faster than a lion to avoid being eaten, just faster than the other antelopes in your herd

The main hope for the industry is that the demand for their commodity will increase, which of course can also be promoted through various activities and will allow more room for producers. In either case, margins for companies in this self propagating cycle tends to be squeezed. This in itself is actually quite beneficial for the overall economy as it creates an abundance of low cost input materials which can be utilised for various purposes.

However, for companies this can be a market economy nightmare, with limited options to influence or escape. They thus have a strong rationale trying to do their outmost to differentiate their product or commodity vis-a-vis their competition, by creating 'after markets' and other services in conjunction with their products. So in general, companies do not want their products to be commodities if they can avoid it—although they not always can.

That being said, companies who have a substantial production advantage in a market with high demand and large fundamental differences in production costs, such as Kuwaiti or Saudi oil producers with pretty healthy margins, escaping the commodity market is probably not something that overshadows their oil companies fundamental strategies.

In general though, decommoditisation lies in the interest of most companies stuck in the bulk trap, and as these companies tend inhabit a substantial part of global economic output, there are strong incentives to push the trend from a business perspective. Part from creating after markets and services around their commodities—investing in information technology to differentiate their products from the rest of the pack will be a logical path forward, since information will be the defining factor in the decommoditisation process. We are thus likely to see a broad range of traditional commodity producers, and commodity users further down stream in the value chain, starting to invest in both marketing resources and information technology expertise to be able to compete in this new environment. Some companies in the paper and pulp industry have already stated it as part of their strategy, and will probably invest significant resources going forward to do just this.

3.5 Nations Push to Escape the Bulk Trap

Part from companies, nations may become weary of falling into the bulk trap as well. This has mainly been the case for lower income countries, some of which seem to been destined to indefinitely produce low value added raw material input to other countries such as cotton, low value metals or crude oil.

In theory there is nothing wrong with this if commodity extraction or production is the countries main competitive advantage. But if the low value added commodity is prone to large swings in price and demand, or is under competitive pressure from an oversupply in other countries; the economic wellbeing of these nations can become very volatile. Since the country's commodity by definition is the same as anybody else's that can produce the defined specifications, and if they cannot form a cartel with the other countries producing the same commodity, then the main way of increasing income will be to increase production volumes or to reduce costs. These actions, which in turn should lower the price of the commodity across the market, will force other nations to either close down uncompetitive production or also go for larger volumes to utilise advantage of scale. Similar situation as the same self propagating cycle companies can find themselves in, which mainly benefits the purchaser of the commodities who are in the position to add value to the end products through differentiation. Just as for companies, the main way of escaping the situation has been to move downstream in the value chain or in some way try to increase global demand.

The bulk trap should not be directly confused with the so called middle income trap, which has similar characteristics, but does not necessary have anything to do with commodity production. A country stuck in the middle income trap implies that they have lost their competitive advantage vis-a-vis more advanced economies in exporting goods, partially due to that slow economic productivity growth is unable to keep up with rising wages. This while at the same time being challenged by low-income economies competing with cheap labour. The growth slowdown usually stems from exhaustion of early development growth pools, such as transferring abundant labour from agriculture to industry and services, technological catch-up or even demographical factors.

This situation slows investment, limits industrial diversification and might also entail that they fail to move downstream in the value chain from pure commodity production. However, in itself it does not have to imply such a case.

Many countries have tried to escape the bulk trap, by differentiating their production base and move downstream in the value chain to increase value added of their exports. For example building refineries and sell petroleum products instead of just exporting crude oil that other nations can refine, distribute, market and sell add on services to the end consumer, which can possibly be more value creating than just extracting the raw resources from the ground. However, the classical economic concept of competitive advantages still holds true—and there might actually be a lot of rationale for countries to just continue specialising in the extraction of resources if this is where they can create most value in the global economy.

Therefore, in order to then escape the bulk trap, countries will need to differentiate and bring value added into the extraction of resources itself. This can be done by attaching information and storylines to their produce which can show that their extracted oil or iron ore is in fact different from that being extracted in many other places due to different factors. The key aspects in this process is information or data, which is transforming most industries today and will probably have an enormous role in commodity production going forward, and might just be what will enable some countries to escape the bulk trap.

Decommoditisation seem to lie both in the interest of the companies supplying the goods, consumers who demand the end products, and even the long term sustainability of our economy and planet. This alignment of interest serves as a strong case of why commodities will face severe pressure to decommoditise, and many of our goods we today see as commodities might go through reproductification journeys in the coming decades. However, there are already some signs on what to come in several markets with substantial externalities evolved. A good example of this has been the energy markets, which previously has been perhaps the most globally commoditised markets, and is currently undergoing a historical information driven transition which we will explore in Chap. 4.

Reference

United Nations Data (2017) World development internet users. http://data.un.org/Data.aspx?d=WDI&f=Indicator_Code%3AIT.NET.USER.P2. Accessed 10 Sept 2017

4 The Energy Market Revolution: A Canary in the Coal Mine

The energy market if treated as a whole, is by far the largest global commodity market in terms of volume and market turnover. However, due to the massive scale in this industry, the externality that its production and distribution causes is arguably also the far greatest. Taking greenhouse gas emissions, global warming and sea acidification into account—the consumption and production externalities of this commodity class may be the greatest threat to civilisation humanity has ever faced. The flip side of the scale of this sector, is that the potential for disruption becomes equally very large, and might be the main reason that this sector is one of those having been furthest impacted by decommoditisation so far.

4.1 Classic Energy Market Logics

Energy has been at the heart of commodities ever since the industrial revolution, and oil is to today the most traded commodity globally.

With the invention of steam engines; combustion material that could feed them have constantly been sought after. It was quickly discovered that the factor that mattered the most was not the tonnes of feedstock, but rather how much heat that feedstock emanated when combusting it. Different sources of wood and coal that were later used have somewhat different heating value per tonne, so the product was partially differentiated by origin.

As the market and sourcing area grew, what mattered was of course not only the material properties of the fuel burned, but also the cost of handling and transportation to the power plant. If you included all these factors into the purchasing price you got a value for cost per units of energy. This unit of this energy has differed across the ages and industries from British Thermal Units (BTU) to Barrels of Oil equivalents (BOE), Megawatt-hours (MWh) or Megajoules (MJ). But common for all is that you measure the energy input regardless of the origin and is really only interested in the cost per energy unit (e.g. EUR/MWh).

F. Andrén-Sandberg, *Commodities Pricing and the Bulk Trap*,
https://doi.org/10.1007/978-3-319-72468-3_4

There are of course some inherent properties of fuels used for combustion that differentiate them, in terms of what boiler that needs to be used, and what conversion efficiency it is possible to get out of your fuel depending on a multitude of factors. Furthermore, with the introduction of liquid fuels and later gaseous fuels it is not possible to treat all fuels as equals per MJ, but rather to compare coal with coal, and oil with oil. Coal power plants have for instance usually lower power conversion efficiency than natural gas-fired power plants, since the latter can make use of gas turbines to produce more power.

However, if having a certain energy conversion factor for a certain power plant as well as the CAPEX and OPEX costs, its possible to calculate the cost of producing that energy, and thus the lowest price you could possibly sell it for. Since the first power sources was rather local, power producers did not really have to compete with power plants in adjacent cities, and prices were thus pretty much set by local conditions.

This all changed with the electrification of industry and introduction of modern power grids. Due to this interconnection of production sources, power producers competed directly against each other, and where judged primary on its ability to produce power at a given moment for as low price as possible. This is only partly true due to bottlenecks in the power grids, but power in the modern area have in general been sold mainly by either energy volume (EUR/MWh) or by capacity (EUR/MW). Either these or a combination of the two, with a complete disregard of where the power actually comes from.

Knowledge Box: Electrification and the Evolution of Power Plants

Electrification and the first central power station was fired up in Surrey England 1881. The system was used to replace gas lighting, and by using +50 arc lamps instead. The system was not a commercial success however, and the town reverted back to using gas lightning.

Nevertheless the trend did eventually gain traction, with central power stations being set up in London and other places in Europe, and in 1890 there where over 1000 central power plants installed globally. Factories also gradually switched from own steam engines to central electrification and household electrification in the US in 1930 was around 70%. Today electrification in OECD countries is more or less completely saturated.

A relatively early invention in the area of power generation was the introduction of CHPs—combined heat and power plants. These did not only produce electricity, but also made use of the excess heat being generated in the process, enabling a much higher conversion efficiency of the fuel. Normal solid fuel power stations today are able convert around 40% of their fuel into electricity, while CHPs achieve an efficiency rate of around 90% by also utilising the waste heat for other purposes such as heating houses or producing steam for industrial processes. Meaning much lower CO_2 emissions per energy produced as well as better economic properties if suitable heating off-takers can be found.

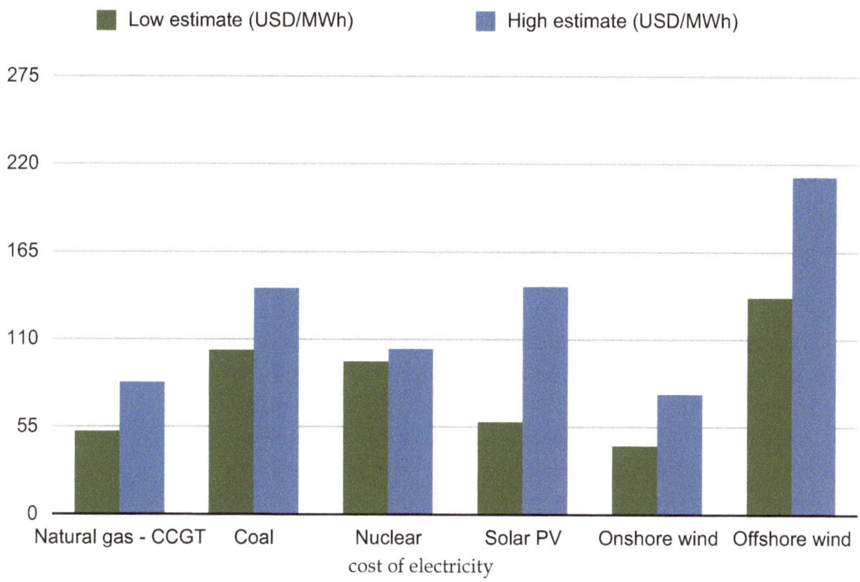

Fig. 4.1 Total system levelized cost of electricity estimation in USA from different sources of power production being commissioned 2022. Source: U.S. Energy Information Administration (2016a, b), Levelized cost of electricity

Since the end costumer could not see any difference in their end product 'electricity' coming out of the socket, they could completely disregard if the power have been produced from oil, hydro, gas or coal—and the only thing the customer needed to care about was the price. This logic implied that the way to make the largest amount of money for producers would of course be to produce and distribute electricity for the lowest price possible, and to compare the cost of energy. The concept of levelized cost of electricity (LCOE) was born.

2016 LCOE estimates for US plants build in 2022 is illustrated in Fig. 4.1—showing large difference within categories depending on local supply conditions. The LCOEs are estimated regarding building and operating new power plants, and should not be confused with marginal costs of producing electricity from plants that have already been built. Should also note that cost for non-commodity driven electricity generation such as offshore wind and solar is decreasing rapidly and that the current costs developments up to 2017 is not reflected in the Fig. 4.1 illustration (Offshore wind LCOE at roughly USD 70/MWh auctioned in Europe 2017).

The LCOE as well as the security of supply has far reaching implications for the international competitiveness for industries, and therefore also for national economic wellbeing. Most countries have therefore implemented comprehensive energy supply plans to gain a competitive edge for their national industries; from nuclear power in France to hydropower in Sweden.

4.2 The Al Gore Moment and the Dawn of the Modern Mainstream Environmental Movement

Historically, there have been environmental concerns against certain energy sources such as open pit coal mining and hydro dams that flood vast areas. But the opposition of these have seldom been coupled with consumers not willing to buy the product that they produce, since there was no way of knowing if the energy was actually produced by a certain power plant or not. Power has thus in general been considered a commodity, and what people cared most about was therefore the price.

This status quo has in recent years has changed though, where people are getting ever more aware of how their power is produced, and where the latest business models enable the consumers to choose their power production sources. The driving force for this progression has primary been due to climate change awareness, and the dramatic information density increase in modern society.

Environmental awareness is nothing new, and has been present during many decades and gained increased traction during the sixties and seventies political movements. The effects of nuclear fallout was a key ingredient in this movement, and several visual oil spills advanced the movement further. In 1972, a United Nations conference on the human environment was held in Stockholm to bring together world leaders to discuss the state of the human environment. The meeting ended with a declaration of 26 principles regarding development and the environment as well as an action plan with 109 recommendations. The principles included: "The Earth's capacity to produce renewable resource must be maintained", "Pollution must not exceed the environment's capacity to clean itself" and "Damaging oceanic pollution must be prevented".

Conservationism was even promoted much prior to this in the nineteenth century, and industrial emissions became a general concern due to smoke emissions and smog during the industrialisation era.

What's certainly new in the current environmental awareness is the inclusion of climate and CO_2 emissions into the agenda, which made previous environmental concerns pale in scale in comparison. The world had become modern through industrialisation, and a key ingredients of the modern industrialisation had been to combust fossil fuel to generate useful energy. In the case of smog, the industry could just install efficient filters to remove the trace pollutants responsible for the problem. In the case of ozone layer destruction, the industry could just replace the ozone harming chemicals with other less destructive ones, and in the case of nuclear fallout from nuclear tests; the world could simply stop detonating nuclear bombs into the atmosphere at beautiful pacific islands paradises!

The CO_2 emissions from fossil combustion is a different beast to tackle, since the fundamental reaction that is responsible for the produced heat; is turning carbon and hydrogen into CO_2 and water in the presence of oxygen, whereas soot and other pollutants usually only constitute marginal amounts and are just biproducts from the main process For instance; the hourly particulate matter partly responsible for smog from a coal-fired power plant can be approximately 10–100 kg per hour, dependent on cleaning technique. The same hour will spew out approximately

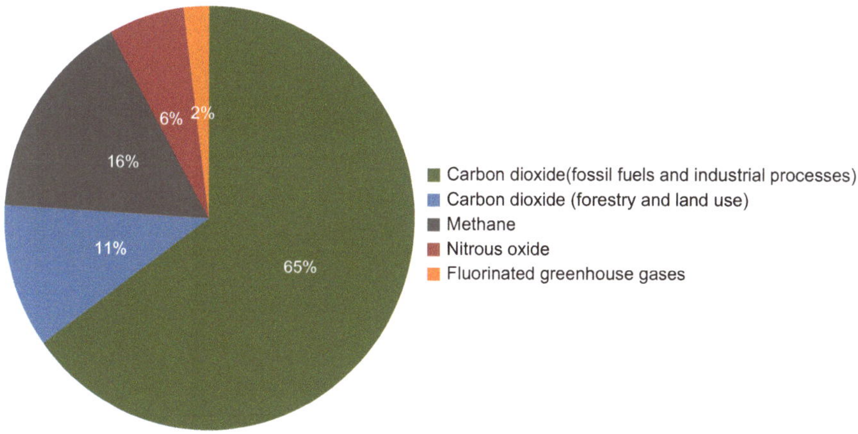

Fig. 4.2 Main sources of global greenhouse gas emissions, as percentage of total. With the primarily emitter being carbon dioxide stemming from combustion of fossil fuel for power and heat generation as well as oil products for transport purposes. Source: Intergovernmental panel on climate change (2014a, b), Main sources of global greenhouse gas emissions

500,000 kg of CO_2 regardless of cleaning technique unless captured—a 5000–50,000-fold factor difference of a compound that is even more difficult to separate out and capture than the particulate matter.

Concern for the impact of the huge amounts of greenhouse gases that were emitted into the atmosphere, and where carbon dioxide was the main culprit, was arguably a fringe worry for most parts of the global population during most of the 20th century. The concern started to gather pace in the 1990s—with the decade ended with the signing of the Kyoto protocol, which albeit modest in it's ambition, was designed to limit emissions from greenhouse gas sources globally. Main sources of greenhouse gases can be seen in Fig. 4.2.

It was not until the early twenty-first century where climate concern became a general concern for the larger population, and the political class. One of many driving forces catalyzing this was the former US Vice President Al Gore, whose documentary "The inconvenient truth" became widely shared online and had a catalytic impact on the mainstreaming of climate concern. This was followed by climate conferences, UN led investigations and even more documentaries as well as research. However, the general impact was unmistakably that the general population had begun to care quite a lot about the climate crisis that were accelerating out of control, and they had become very keen on doing something about it!

4.2.1 Power Markets Fundamentals Turned on It's Head: The End of LCOE

Companies and governments has been taking note of this trend, and responding with the build-out of renewables as the most visible sign of this shift that has swept

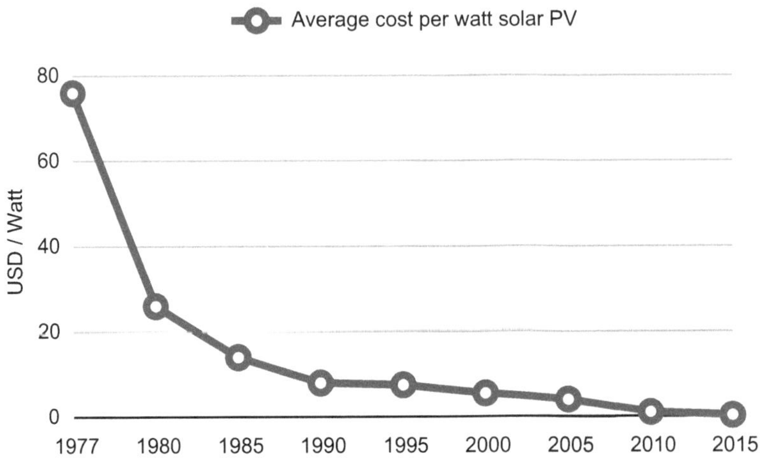

Fig. 4.3 Solar panels production costs. Which have seen a dramatic 200-fold price decrease in the last 40 years, and is starting to become competitive with conventional power production within several geographies. Source: Bloomberg New Energy Finance (2016), Evolution of PV solar panel cost

the globe. Renewables power generation has generally been more costly than their fossil alternatives, which is the main reason they've not been built out in the first place. To promote renewable energy growth to replace fossil generation, many governments around the globe have begun to tax fossil fuels while subsidising renewables. What this actually implies is that nations have started to value some sources of electricity higher than others, and are willing to incentivise renewables and penalise fossil fuels in order to promote a shift in primary energy generation. Which in itself can be seen as partial decommoditisation of energy, it nations are considered as the purchaser.

The main buildout of renewable power in the world in addition to hydropower, which historically has been competitive in its own right and build-out wherever sufficient conditions existed, have been biomass, wind and solar energy. The solar energy trend that has swept the globe have in particular been quite astonishing and is good example of how this shifting trend has impacted and progressed in the market.

What mainly started out as a German incentive to phase out their coal and later on nuclear power capacity, drove down prices on Solar Photo Voltaic (PV) power panels in such massive way, that other nations saw it as an increasingly compelling option to follow suit. Cost development for Solar PV can been seen in Fig. 4.3.

Large part of the solar PV build-out has been based on subsidy regimes that does not only give large scale power plants sufficient remuneration, but also incentives for ordinary people in order to install solar panels on their roof or other areas adjacent to their home. Thus creating a decentralised power generation system.

Another interesting part of the trend is how regular people and companies are commissioning small scale solar production, even when there's doubtful economic rationale for those installation. The main reason seemingly that the value of producing your own electricity may be higher than simply buying it of the grid, just in the same way as tomatoes grown in peoples own yards may have a higher value for some consumers than those bought in the store. It may thus be more valuable to have solar panels on your house to show the neighbours your green credentials, or for companies to show the customers the same thing than buying regular or solar energy from the power grid.

Knowledge Box: Solar Energy Facts

- The amount of solar radiation hitting earth every hour, is more than enough to power all human energy requirements for a whole year. However, since the efficiency of solar panels is around 16% it would take a few hours for a globe covered with panels to be sufficient of all humans yearly energy need.
- To generate enough power to supply the global annual energy needs with only solar PV, you will need to cover approximately 0.3% of the earth total land area depending on where you put them—or less than 4% of the saharan desert.
- However, these amounts would only solve the total volumes of power needed, another issue is capacity sufficiency where modern economies have a constant need for power in our grids. This would need to be solved through storage or energy conversion of some sort e.g. Pumped hydro or Power-to-gas storage in order for full renewable penetration to take place.

This is not to be disregarded—if some of the cost of installing power production can be attributed to creating goodwill, then the LCOE of the technology could actually decrease substantially when only part of the cost is attributed to power production. This shows that production of power is not any more as easy as comparing levelled cost of electricity, and the commodity of power can hardly be considered to be a pure commodity anymore.

A somewhat more abstract form of production control is the financial purchase of whole or part of wind mills by individuals and other non-utility entities. This has become rather popular with companies who want to prove their green credentials; showing that they care about the planet and are working towards a sustainable future, which might give them an advantage vis-a-vis competitors. An example of this is the furniture company IKEA, which has acquired wind mills and installed solar panels on many of their warehouses. With the goal of producing at least as much power from green sources than it consumes by 2020, thus being a net neutral consumer of external energy on a yearly basis.

Another example of the decommoditisation of electricity is that consumers in some markets now can choose from which production source they wish to buy their power from; with alternative sources such as wind, hydro or just plain electricity mix. The drivers for consumers picking some means of production if favour of other is primary due to that transparency in the market have increased dramatically with the progression of information technology, and consumers are now much more informed about the externalities involved.

Fossil power plants and their fuel is considered vital infrastructure and getting access to the sites is usually restricted, and they're also hidden away in some sense, as large scale lignite and hard coal production are kept well away from densely populated areas. Due to this, the primary tool for obtaining and spreading awareness in the past was trough books and documentaries. But with the trend of ever increasing information transparency and with the progression of the internet in general an social media in particular, the information of production methods have become ever more proliferated. Consumers are not only better informed of what choices that are possible to make, but they can now also share these choices to their peers and thus increase the value of their purchase; perhaps enhancing their social status by showing of their own green credentials.

However, just to be clear—the notion that consumers can purchase a certain set of electrons differentiated from other electrons in the power socket is more or less absurd. Unless the customer are situated in an island grid with only one production source (which is very rare), the electricity in the socket will almost always be produced by a blend of power sources who does it best to keep the frequency in the grid at a certain rate, and it will be impossible to actually get the power from a certain source in the grid.

Nevertheless, regulators and the power industry have solved this 'absurdity' by setting up renewable certificate systems, in which 'virtual' certificates are created whenever renewable power is produced. These certificates can then be sold onwards to consumers in conjunction with the regular power, which raises the total price of the renewable electricity and gives producers of renewable electricity additional incentives to produce.

This system is pretty clever since it gives an incentive to those production sources that consumers demand and would actually like to see in their power system, by creating a price premium for renewable producers. An interesting note is that the certificates themselves can usually be traded, where the price are set on a market based on supply and demand—and has thus itself become a commodity.

Due to this support and massive build-out of renewables, the once very expensive options deployed for altruistic reasons has now become competitive with their conventional peers on an LCOE basis in many areas of the globe. As of 2017—onshore wind energy in Texas is among the cheapest forms of new-build energy that can be built globally. Offshore wind costs in Europe had been reduced by 50% in just under 3 years, and solar energy deployed in the Dubai and Chile have been able to produce power from sunshine which costs roughly half of the coal generated power alternative. According to Bloomberg New Energy Finance publicly available news data—this trend is also forecasted to continue, with the global average cost for large scale solar parks expected to drop another 50% from 2017 to 2025 (from a 2009 baseline the drop to 2025 is 84%).

The reason for this is the key important difference between technologies and commodities. Commodities tend to get more expensive as demand increases, due to limited supply and even exhaustible supply in the long run. For example, if demand for oil increases, more expensive fringe sources needs to be taken into production, thus raising the price for the commodity as a whole since governed by marginal price setting.

Technologies on the other hand operates in a very different manner, since they tend to get cheaper the more they are deployed due to the so called learning curve, or learning effect. If the learning factor is say 20% (such as in the case for solar panels), a doubling of output will bring down the cost by 20%. So if an increased amount of solar panels are being deployed, the prices will be reduced for the next purchaser of solar panels, which will incentive even further deployment of solar panels, which will lower the price for new solar panels, which will incentivise even more massive deployment and so forth.

This is exactly what has happened in the energy industry with both solar and wind power, which has now both transpassed the threshold limit where they through initial support are now able to compete in power markets without subsidies in many areas of the globe, due to competitiveness vis-a-vis coal, nuclear and natural gas. And since they are now underpinned by fundamental market forces, there is an entirely different momentum driving the industry going forward which will probable reduce prices even further.

Taking these developments in consideration when looking at how the power market has and is evolving, even though costs matter and the build-out has even accelerated by ever cheaper renewables—the levelized cost of electricity does not anymore seem to give an accurate description of what drives the build-out of new power production. Implying that the power market which has been on of the most commoditised markets—has now been partly decommoditised in the eyes of governments, companies and end consumers.

4.3 The Drawback of Intermittently, the Obstacles for Extensive Renewable Build-out, and Why LCOE Do Not Tell the Whole Picture

One of the drawbacks of most renewables is that the energy production of for instance wind, solar or wave energy is largely dependent on the energy density of the site, as well as the area harvested. Compared with a nuclear power plant which may occupy a large industrial site, a solar energy or wind farm needs to take up at least 100-fold the surface areas of a nuclear power plant to be able to produce the same amount of electricity. Furthermore, the renewable energy source usually needs to be deployed, not where its mostly needed from a demand perspective, but rather where the energy density for wind or solar is high enough or where it disturbs the least amounts of people. The latter has proven to be a major obstacle especially for onshore wind in densely populated countries, where it has become increasingly difficult to find sites with decent wind speeds without complaining neighbours.

That being said, the total human energy demand could easily be supplied from a geographical standpoint with a combination of solar, wind and other renewable alternatives if sufficient public acceptance can be achieved. Covering just under 4% of the Sahara desert with solar panels would for instance be sufficient to power the entire global energy need on an annual basis.

The main drawbacks of most renewable energy sources is rather that they are so called intermittent source of energy—where the sun is not always shining exactly when the consumers wish to turn on their appliances, and the wind cannot be directed to blow evenly at all hours of the year. This is pretty substantial problem since electricity has an interesting property that not many other goods or commodities have—that it needs to be produced at the exact point in time that its being consumed or just a few milliseconds before.

Most power networks operate with AC (alternating currents) technology where power is being distributed through maintaining the exact same frequency in the grid, such as 50 or 60 Hz (this too standardised from a plethora of frequencies even within countries during the last century). If some consumer uses power, this causes the frequency to slightly drop, which must then be compensated by a producer putting in more power into the network which raises the frequency. The role for the so called TSO (transmission system operator) in most countries is to keep the frequency around a firm number such as 50 Hz, with a confidence interval of perhaps 49.9–50.1 Hz. If the frequency where to drop further down to 30 Hz or similar, a lot of our appliances such as computers, electric stoves and TV monitors would simply start to break down, the same applies for industrial machinery which might even be damaged in the process.

To traditionally keep the frequency stable, the TSO could just dispatch a certain number of power plants in any given hour to be able to satisfy demand. For example running ten nuclear reactors and four coal-fired power plants at all times, and in the evening peak hour when consumers came home from work or in the event of the Football World Cup TV finals—they could dispatch another gas-fired power plant for a few hours to keep up with the increased power demand.

With intermittent energy sources, the situation changes quite a bit since the TSO cannot choose when the wind turbines should deliver their energy, or when the sun decides to shines on the solar panels. Instead they will have to try to balance the intermittent renewables by dispatching other power plants when the renewables are not delivering, and the system must therefor have a lot of spare capacity available in order to balance the supply and demand. However, the problem with this is twofold. The spare capacity which is dispatchable, will not run as many hours of the year as they did in the past when they were the main base-load capacity. Thus, to be able to retain profitability and cover their investment costs base; they need to receive higher remuneration per unit of energy when they actually produce power. The other side of the coin is that when the renewable energy capacity reaches a certain limit on their way of supplying the whole energy system with their power, the price

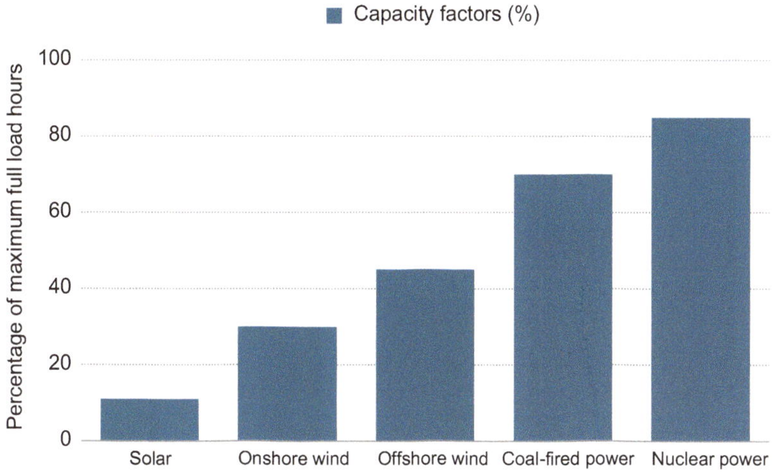

Fig. 4.4 Capacity factors for select technologies. 100 MW of solar PV will according to this example produce on average produce 11 MW of power on a yearly basis since there is no sunlight shining during the night and only scant sunlight during the winter months and cloudy days. 100 MW of US nuclear power stations would on average produce 85% of the time and thus averaging 85 MW on a yearly basis. Sources: Swedish Energy Agency (2017), Energy characteristics and U.S. Energy Information Administration (2017), Capacity factors

they are able get from the power market will start to diminish. The reason for this is that the producing capacity will actually approach or even exceed demand at certain hours—and when supply exceeds demand the price moves towards zero. Thus even if renewable energy are coming down in cost, the production profile causes substantial problem in itself since the captured revenue moves below the average power price, while revenue for conventional generators increases above the average power price, hence counteracting the competitiveness of renewables. Capacity factors for select energy sources is illustrated in Fig. 4.4—showing the percentage of of full load hours per year of an average utility scale installation. The renewables example are based on Scandinavian latitudes, while US base load is assumed for coal-fired and nuclear power stations.

US based nuclear power stations produces full power roughly around 85% of the time, in theory implying that in order to supply enough energy volumes on a yearly basis a nuclear capacity $1/0.85 \approx 1.2$ times the average demand must be erected if only utilising nuclear power. Solar PV (PhotoVoltaic) panels produces approximately 11% of all hours in Scandinavia, meaning that the Scandinavians will need solar capacity $1/0.11 \approx 9.1$ times the average demand to meet the yearly energy needs. However, the problem is that meeting the yearly energy demand is simply not sufficient since the consumption load is fluctuating quite a bit. Power producers therefor needs to supply not only average demand, but also peak demand. If the power system capacity was capped at average demand, the grid area would experience partial power shortages and blackout approximately 50% of all hours somewhere in the grid. This principle being illustrated in Fig. 4.5. The energy

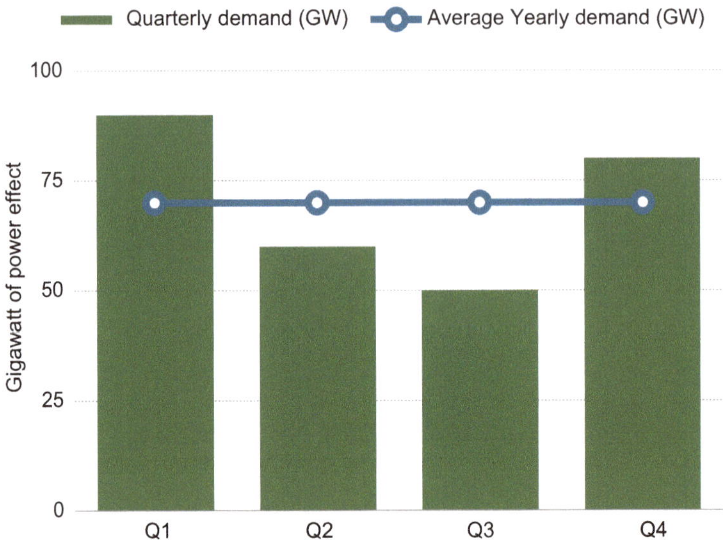

Fig. 4.5 Illustrative power demand per quarter and average over a year. Showing that if only supplying capacity to satisfy average yearly demand, the system would only have enough power during Q2 and Q3, with some over capacity in those months, while experiencing power shortages in Q1 and Q4

system rather needs to be tailored in order to be able to supply dispatchable energy sources at the peak demand hours every year, including extreme demand years in order to avoid blackouts.

The problem is that renewable intermittent energies will not be able to supply their power exactly when consumers need it. So if supplying the energy system with only solar to meet yearly energy demands in Scandinavia, the production would probably exceed demand tenfold in the summer months—while supplying a tenth of the needed power in the cold winter months due to seasonality in demand and solar influx.

Due to this there is the need to move energy volumes in time to another hour when the sun is not shining or the wind is not blowing. There are basically two ways to do this; either you make sure that you have sufficient backup capacity to supply power at all times and combine this with sufficient grid build-out, or you deploy large amounts of energy storage.

If there's sufficient power grid build-out, it's possible to get rid of the excess power to someone else who needs it as that hour—and import power when the conditions are reversed. This can however be bit problematic for solar producers in some regions when most producers would get their sunshine at the same time. The power transmission lines would in that case need to span quite some distance and be of pretty thick caliber to be able to handle the peak production volumes. It would neither solve the issue of lack of sunlight during nighttime, unless the power lines were extended over several times zones. If they were, it would incur substantial electricity losses in the grid and more power would thus had to be built out, increasing the overall cost to the system. That being said, increased grid build-out would help alleviate solar flux over large areas, but would not be sufficient in itself.

However, grid build-out suits very well to for instance balance wind power deployment, where over a continent there is almost always wind blowing somewhere, and demand arising somewhere. Another way of solving the intermittency problem though, is through energy storage.

You can't direct the wind—but you can adjust your sails

Energy storage was not seen as a major part of the energy system in the past but have received a substantial upswing lately with the build-out of intermittent energy sources. The rationale for energy storage is that you can simply store the energy produced from the wind or solar resources for later usage. The main storage that has been deployed up until 2017 globally has been pumped hydro reservoirs or simply hydro reservoirs. Where in case of the former; water is pumped into a dam when there is excess power available in the grid, or in the case of the latter; water is just saved in a dam when not needed for later usage.

This is a great way to balance the power supply, and can also be done rather cheap. The obvious drawback is that it can only be done wherever hydro sources and dams are present, and in developed economies most the build-out potential has usually already been utilised. Although the current hydro capacity was sufficient to balance an energy system that relied mostly on dispatchable energy sources, it will in most countries be insufficient to fully balance a switch to a complete intermittent renewable energy system. Which is a pledge that many countries and regions have vowed to implement within the coming decades.

The main energy storage candidate to balance renewables and complement hydro so far has been through regular large scale batteries, which can be deployed in sufficient capacities to relieve congestion and balance intermittency throughout the day. The main drawback of batteries is that the energy volumes that can be stored is typically rather small. A battery of 1 MW which costs approximately EUR 2–3 million in 2016, can usually only store around 4–8 hourly times it's own capacity (4–8 MWh), roughly what a single European house uses on a yearly basis. This makes batteries useful for intraday usage, by storing solar energy from the day and be used throughout the night in areas which has large and stable amounts of solar influx, or to balance wind output over a few hours. However, it unfortunately makes batteries unfeasible to store large amounts of energy over seasonal time spans, from for instance from the summer to the winter. This would be needed if using solar energy at a large scale in seasonal climates such as in northern Europe and US, parts of China and Japan. The additional cost would simply be too large to become practical, even if the prices for batteries were reduced to a tenth of todays costs.

Knowledge Box: Power-to-Gas for a Sustainable Energy System

To achieve tomorrow's sustainable energy system, there needs to be a large supply of renewable energy coming into the system from sources such as wind and solar. However, this will unfortunately not be enough. Since the wind does not blow and the sun does not shine at all times, there needs to be some way of storing the energy to whenever we want to use it. Ordinary batteries is an obvious option, but can only offer intraday storage with current technologies (i.e. from the day to the night), which is insufficient for countries in the global north with a seasonal energy imbalance, and thus a seasonal energy storage need. Another option is utilisation of hydro power reservoirs, which is currently widely used for storage purposes. Although these are unfortunately only available in certain geographies, and are not in such a scale that it can support an entire global energy system based on intermittent renewables.

However, a viable option that can be applied at most places is storage through Power-to-gas (PTG/P2G). In this process electrical power is tuned into chemical energy by spitting water molecules into its constituents oxygen and hydrogen. The oxygen is released in to the air, while the hydrogen coming out can either be stored directly or made to react with CO_2 to form methane or other molecules to form even more easily stored compounds. The produced gas can then be stored inter-seasonally from the summer months when energy demand is low, and used to produce heat and power in the wintertime when renewable power production are insufficient to meet the elevated demand. The storage of the gas can be integrated in currently established natural gas infrastructure, or at dedicated storage sites depending on if the gas is in the form of hydrogen or methane. The produced gas can also be used to fuel gas powered vehicles, thus decarbonising long distance transport where electrical battery vehicles might be less competitive. Some of the initial electrical energy is, however, lost in the conversion processes in the form of heat, but can also be utilised if cleverly integrated into the energy system to provide heating where needed.

Power-to-gas is a well proven technology, with a declining costs curve of deployment. The drawback though is that it still costs more to produce the gas, than the current fossil natural gas option. However, this assumes that the conventional natural gas does not have to pay for all the negative externalities (such as CO_2 emissions) it causes. If these externalities these where to be priced in, power-to-gas might well become an integral piece in achieving the sustainable energy system of the future. It is in any case difficult to see any other viable options which actually has the potential to achieve an energy system that can produce the required energy at all times without an integrated power-to-gas solution. The really good news is that there actually is this viable option available to integrate solar and wind energy in our energy system, and taking civilisation perspective factoring in global warming, it arguably looks like a quite affordable solution as well.

For seasonal climates there are other candidates to store energy such as power-to-gas and others that might offer a way out the issue, but the general takeaway is that the LCOE is simply not the only thing that matters anymore for competitive power generation. Production profile in relation to demand and storage cost, grid build out and backup capacity must now also be taken into account. In addition to this, there is the public acceptance part of the equation where wind power is seen by some to negatively affect the landscape picture, and some regions have even banned the deployment of onshore wind turbines to preserve their citizens balcony view. Solar can perhaps be deployed rather inexpensively with less public friction, but has the drawback that it will never produce in the nighttime, and neither in the wintertime at northern latitudes.

The solution going forward will probably be a combination of massive renewable build-out in conjunction with substantial new grid build-out, storage and backup capacity—enabling system operators to keep the lights on uninterrupted at all times. These technical aspects furthermore needs to be solved while at the same time listening further to what the end costumers really want, and not only focus on pure commoditised power production.

An interesting point in the power sector storage evolution, is that trend for increased deployment of batteries has actually not been made possible from mainly power suppliers—but rather from the transport sector which in itself is on the verge of undergoing a substantial evolutionary journey.

4.4 The Seeds of a Transportation Revolution

Another energy pillar that has been impacted by the increased public pressure is the transportation sector. Being the main consumer of oil products both through usage of fuels such gasoline and diesel, but also in terms of asphalt and other refinery fractions—the sector has been responsible for a substantial part of global CO_2 emissions.

Part from CO_2, compared to large fossil fuels-fired power stations which can 'clean' their exhaust in various ways to reduce non-CO_2 pollutants, cars are just simply too small and distributed to be fitted with similar efficient cleaning technologies in an economical manner. These non-CO_2 pollutants such as nitrogen oxides, sulphur and particle emissions which can be damaging to public health and the environment has come under increased public pressure due to dangerous levels of smog in many cities across the globe. The increased health hazard has even urged authorities to limit or even ban diesel vehicles in several cities across the globe. Although utilisation of diesel may entail lower CO_2 emissions than petrol, due to the inherent property of the blend and its combustion, it has proven difficult to reduce small particle emissions which causes local harm to human health.

Cars also make a lot of noise, takes possession of large parts of the cities with parking lots and roads, and causes geopolitical strain where oil is one of the most important commodities traded globally. There are signs that many of these externalities could be about to be mitigated though, with the introduction of the electric car and new types of unconventional fuels.

Electric vehicles have been around for a while and was actually in close competition with combustion engines in cars in the early twentieth century, with roughly 38% market share in the US compared to 22% gasoline market share (the rest was steam powered). Combustion engine cars did however had the advantage of substantially longer operating distances per weight, and they were also quicker to refill with cheap fuel stemming from oil. With the advent of ever better road networks outside the cities, the combustion car overtook the electric car due to its superior range and has become humanity's main choice of vehicle ever since. Electric vehicles have been reintroduced now and then since its decline, but were treated mostly as a curiosity and did not catch up with the general consumer. It also had bit of a 'geek' status, that was deemed hard to penetrate.

This may have changed irreversibly when the company Tesla introduced their quick and high status car in 2008. Their Tesla Roadster was as sleek and quick as any conventional sport cars without did producing any CO_2 emissions, nor made any significant noise. Some years prior to this, the semi-electric Prius had been introduced and gained significant popularity as it reduced carbon emissions through recovering break energy from the vehicle. But with the introduction of the Tesla; the entire perception of what an electric vehicle was arguably changed.

The initial Tesla did not make a great impact on the overall market in terms of sold volumes, but the electric vehicle concept seemed to possess a significant add on status value; and soon other car manufacturers such as BMW, GM and Mitsubishi followed suit with their own electric vehicles lineup. Although the sales figures as far as 2017 for electric vehicles were not anywhere near conventional cars sales, having sold approximately 780,000 vehicles globally through 30 models in 2016, the sales growth is what has astonished industry analysts—with a global 40% YoY increase in 2016 and 50% increase in European sales during 2017. The main trend shift being that in just a few years time; the electric car had gone from a 'geek toy', to a high status vehicle type that could appeal to the masses.

Owning an electric vehicle showed that you cared about the environment, and once the car was bought it furthermore lowered the daily commuting costs with roughly 80–90% from a pure fuel perspective. If including the degradation of the battery and the initial higher costs; the picture would be somewhat different and would probably entail a slight loss, but people did not seem to care. The main reason to buy an electric vehicle was not to reduce costs, but to be able to do something tangible for the environment yourself and arguably to be able to show your peers that you did so as well. Caring for the climate had in many social circles became desirable, and those who could afford where happy to become early adopters.

The development also offered a path for governments to reduce fossil oil dependence by subsidising the purchase of electric vehicles, which were given tax breaks and different rounds of incentives in many countries across the globe.

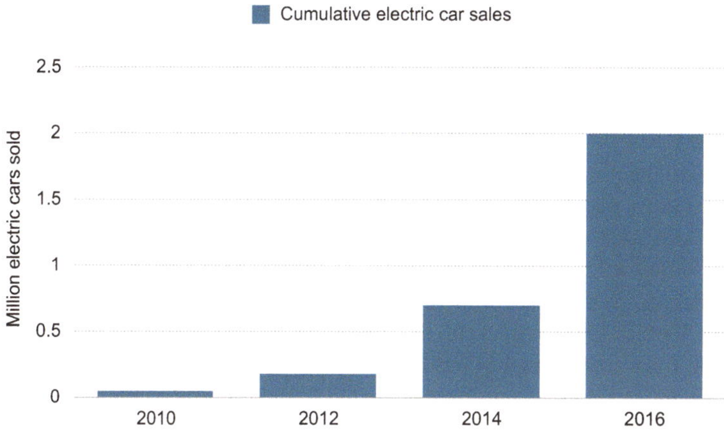

Fig. 4.6 Cumulative global electric and plug in hybrid car stock. Source: © OECD/IEA (2017) Global EV Outlook, IEA Publishing. Licence: www.iea.org/t&c; as modified by Springer

Incentives which would not had been feasible if not significant parts of the general population were positive towards electric vehicles.

Although electric vehicles had as of 2017 not penetrated the overall car market in any substantial way yet; with less than 1% of global overall car sales. A sign of things to come might though be seen in the progress in Norway—where the market share of sold electric and hybrids vehicles reached over 30% in 2016. The significant growth of the industry can also be seen in Fig. 4.6, with estimations for future EV car stock ranges from 25 million to 70 million vehicles in 2025. Quite high numbers that have a fair chance of being surpassed judging the underestimations of former forecasts in the recent energy trend.

The incumbent car industry have noticed this trend as well, and most car manufacturers as of 2017 have electric vehicles either in their offerings or product pipelines. Volvo announced in 2017 that all vehicles beyond 2019 would be fitted with electric motors, BMW has announced electrification of all the group brands and model series, and other main manufacturers have similar programs for their pipelines. Furthermore both France and the UK had put forward regulation that forbids sales of diesel and petrol cars beyond 2040. Regulation which interestingly enough the industry analysts did not attach too much weight to, since pure market fundamentals estimates that electric vehicles will be highly competitive well before that.

What had allowed the Electric Vehicle (EV) revolution in first place was due to mass production of batteries for the computer and electronics industries. Lithium ion batteries which offered higher energy to weight performance compared to alternatives, had due to the increased production volumes become substantially cheaper, and thus overcome the threshold where it began to make sense to use in vehicles. With the electric vehicles industry ramping up volumes and demand for batteries from the electronics industry showing no signs of abating, prices for batteries and thus for electric vehicles are expected to decrease even further in cost going forward. As the trend continues, there will probably be a moment in the coming decade where electric vehicles has become cheaper than their petrol cousins, which should be the tipping point where the market shift towards majority of EV sales.

Meanwhile petrol powered cars might conversely become more expensive in the process as their sales volumes are being reduced. Or to put it in another way, the reason petrol cars can be competitive today with electric vehicles is that they currently enjoy so much larger production volumes and thus efficiency of scale.

The electric vehicle revolution does not have to be the only game in town however. Part from electric vehicles, there has been a flurry of activities on the renewable fuels front. Ethanol production and sales have been subsidised and sold in large quantities both in Europe and the USA, the liquid also fuels a major part of Brazils vehicle fleet.

Biodiesel have been mandated to blend into conventional diesel in several countries through a rising quota system, and there are even 100% blends available in the market produced from different types of feedstocks. Biogas and natural gas has also been promoted to vehicle customers in many countries, as a mean to reduce car emissions.

Some companies are also starting to offer hydrogen powered vehicles, which offers substantially shorter refuelling times compared to electric vehicle peers, while entailing similar characteristics in regards to carbon emissions and other pollutants.

Also flight transport are beginning to see modest change with bio based fuels being tested by several companies in order to reduce CO_2 emissions. Even electric flight for smaller planes and shorter distances are being explored, and are getting a boost from the booming drone industry who's possible future impact on transport and delivery of goods should not be underestimated.

Ocean shipping and sea transport are also beginning to slowly change, with several regular ferry lines in the progress of electrification and long distance shipping beginning to use and view liquified natural gas as the fuel for the future instead of bunker oil. The switch from bunker oil to natural gas in shipping, and in the long run even biogas or hydrogen, would not only have a significant effect on CO_2 emissions, but also on local air pollution which has shown to negatively affect the health of populations living close to harbour areas. The cruise line Royal Caribbean has for instance announced it will start using hydrogen fuel in the future to enable zero-emission propulsion depending on how the hydrogen gas is made. Global shipping and delivery firms has also started to show their engagement, with cars running on renewable fuels and statements regarding being clearly communicated on the side of the cars.

The whole transport revolution which is beginning to gather pace would probably be an own book in its own right—but the main takeaway is that transport alone is not the only concern anymore. People and governments are starting to internalise the externalities invoiced in transport, and are starting to demand that their shipping is done with less air pollution, that their Caribbean cruise does not harm their environment (and thus their conscience when they are sipping umbrella drinks on the luxury liner's balcony). Moreover, many certainly want an alternative to petrol-driven cars in their daily life which does not harm the environment or their fellow neighbours with local air pollution.

The revolution will pose entirely new challenges and opportunities for the energy infrastructure, such as hydrogen production facilities or electric charging infrastructure. Coupled with increasingly sophisticated artificial intelligence software, the trend will probably change the transportation system in ways that has yet to be seen or even imagined. The one thing that does seem certain is that transport and the fuels powering will probably not be regarded as just another commodity by the consumers of the future.

4.5 The Evolving Gas Market

Natural gas stands right next to coal and oil as the most used form over fossil energy globally. The main reason being its relative abundance and the versatility of usage, with can stretch from simple heating by combustion, power generation or used as a chemical feedstock in industrial processes.

The main component of natural gas is methane, which is arguably the simplest organic molecule available with just a single carbon atom surrounded by four hydrogen atoms CH_4. The simplicity of this molecule is also what makes it so abundant, since the molecule is produced in many chemical and natural biological processes and can essentially be created from most other organic compounds available. When organic material responsible for the deposits of oil and coal is imbedded in the ground without any oxygen available to break it down into CO_2, parts of the organic fraction starts degrading albeit slowly over time into methane, which is why natural gas is usually found in conjunction with other fossil deposits.

When the first oil wells where drilled, natural gas was seen as a nuisance and even a hazard—since it was a highly combustable gas coming out of the ground whenever oil was being recovered. Due to the risk of letting the combustable gas just linger around in the area, many oil drillers decided to just combust or so called 'flare' the gas. However, as demand for primary energy increased and prices for crude oil started to rise, producers soon realised that the natural gas could be used as an energy source as well. Today, natural gas is a major contributor to the global energy supply, with its share being illustrated in Fig. 4.7.

Knowledge Box: The Relative Scale of Fossil Fuels and Renewables

Even though fossil energy is by far the largest contributor of world primary energy supplies, supplying approximately 80% of primary energy supply, it should be noted that its still a relative small amount in a planetary context. Solar radiation on earth on sea level is approx. 7,500,000,000 TWh per year,

(continued)

or 47,000 times more solar energy than the entire annual global energy demand. Hence, its very much possible to supply total energy need from renewable sources such as solar and wind energy if humanity decided to go down such a path.

A positive feature of natural gas is its relative clean combustion, as it doesn't emit much sulphur or other harmful particles compared to coal and oil derivatives. When used in power plants, natural gas can also achieve very short startup times to reach its full power capacity due to the plant design. Which means that it can respond to shifts in demand very quickly, and is thus suitable to balance intermittent renewable power sources.

However, the main usage of natural gas has been to provide heating in residential and commercial spaces—where it for instance in the EU is the most common form of heating covering approximately 40% of overall heating demand (heating itself responsible for approx 50% of overall energy use in the EU!). Natural gas also has the convenient property that it can be distributed relatively easily across continents through pipelines in contrast to bulky solid fuels, or across oceans in liquified natural gas supertankers.

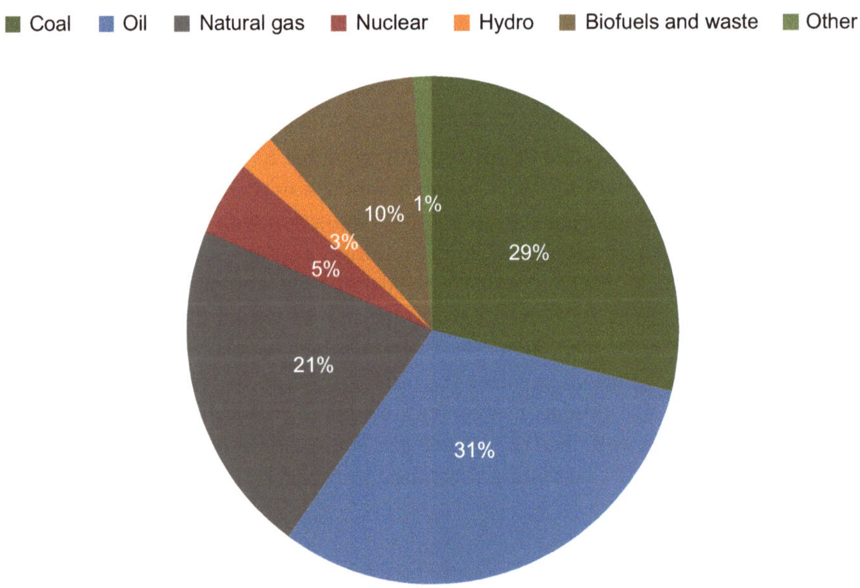

Fig. 4.7 Total world energy primary energy supply by energy source. Natural gas supplying approximately 20% of total volumes, or four times as much as all nuclear energy to put it into context. Source: Intergovernmental panel on climate change (2014a, b), World energy supply

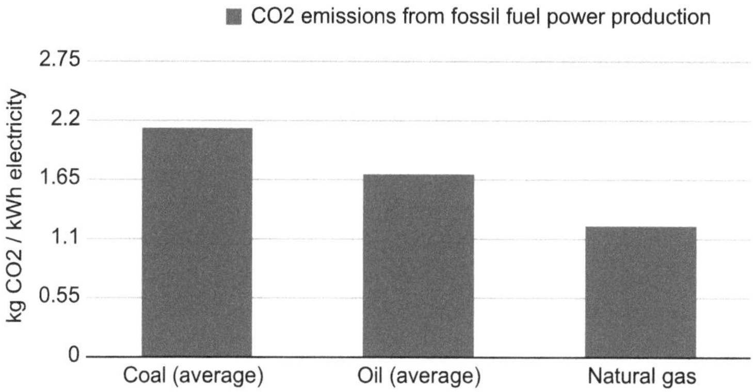

Fig. 4.8 CO_2 emissions from range of fossil power production alternatives per produced unit of power. Natural gas-fired power stations usually emit significantly less CO_2 emissions per produced unit of electricity compared to other fossil fuels alternatives. Source: U.S. Energy Information Administration (2016a, b), Carbon dioxide emissions from power generation

The drawback of natural gas is however, that albeit it's a relative clean energy source and delivers more units of energy per emitted CO_2 than other fossil fuels as seen in Fig. 4.8, it does still add net contributions of CO_2 into the atmosphere and thus drives climate change. Also since it's a fossil fuel and not a renewable one, it furthermore have the obvious drawback that the resource will eventually be depleted. The good news is that it's possible to derive methane from renewable sources as well, where the net contributions to CO_2 in the atmosphere is usually much less depending on production method.

Biogas, which is the most common source of renewable methane, is produced by letting bacterias devour digestible organic matter in an oxygen free environment. Since there's no oxygen present, the bacterias cannot digest the organic matter into CO_2, and instead resort to produce methane as a byproduct of their digestion. This process is currently being managed at several sewage treatment plants, where human waste is put into large oxygen free chambers and allowed to 'ferment' while methane is simply collected at the top of the chamber roof. Another example is dedicated anaerobic digesters, where everything from manure to food crops are put through the same process, with the sole purpose of producing renewable methane.

Another less common way to produce renewable methane is through gasification of biomass, where feedstock such as woody biomass or straw is heated to elevated temperatures in an oxygen scarce environment—turning some of the longer chemical carbon chains in the material into gaseous methane.

All these above methanisation processes can also be enhanced by power-to-gas technologies, where electrical power is converted into chemical energy through electrolysis of water. This process runs by splitting water molecules into its constituents oxygen and hydrogen, where the latter is afterwards made to react with CO_2 to produce water and methane CH_4.

These processes usually produces renewable methane with much lower net carbon emissions compared to the extraction and combustion of natural gas. Although, the interesting part is perhaps not the methods themselves, but rather that the production methods as such is starting matter for governments and end consumers.

Natural gas is by far the cheapest source of methane, and is usually taxed by government to produce an income for the state. Renewable sources of methane on the other hand is by comparison fairly expensive to produce, and usually needs to be subsidised by governments in order for it to be produced. This is exactly what is starting to happen across the globe, where governments are deploying different subsidy systems to incentivise the production of renewable methane—either to be combusted directly to generate renewable power and heat, or injected into a natural gas grid to increase renewable gas supplies.

The problem in the latter case is that the renewable gas when injected into the natural gas grid is impossible to differentiate from the general natural gas at the source of usage—it becomes a blend. On the other hand, it would be very expensive to build up separate distribution systems to differentiate the renewable methane gas from its fossil cousin.

The way this is usually solved is in a similar fashion to renewable electricity—to provide green gas certificates for the renewable gas that being produced and injected into the conventional natural gas grids. This certificates can then be sold onwards to end customers in conjunction with the methane, ensuring that all biogas volumes sold is also being produced somewhere. The end consumer may in many cases be burning natural gas molecules, but since ensuring that similar amounts of biogas has been produced somewhere, the environmental benefit is the same as burning biogas directly by crowding out similar amounts of natural gas volumes in the system.

In addition of being better for the climate, the renewable methane also ensures energy security since it can be produced domestically, in contrast to imports from nations which may use natural gas as a geopolitical bargaining tool. It also entails the added benefit that it may create local job opportunities, and might increase competitiveness of domestic agricultural and food industries by adding revenue streams compared to international competitors.

Knowledge Box: Low Tech Biogas

Biogas production is a fairy easy process and could even be done at home by relative simple means. It essentially putting organic biomass that can be digested by bacteria in a oxygen free environment, such as a covered hole in the ground, and collect the biogas being produced. Such simple biogas reactors have actually been deployed in countries such as India and China, where human waste, manure and food waste is put in a tank or a sealed hole in the ground covered with plastics. A tube is then connected to the top of the

(continued)

chamber so that the gas can be transported to wherever it's needed, such as directly into the kitchen for cooking. This being a cheap and efficient solution to improve rural sanitation and ease pressure on firewood resources.

The technique may have been used by earlier civilisations as well, with Marko Polo reported that the Chinese extracted energy from covered sewage pots back in the thirteenth century, and some historians even claim that the Assyrians used biogas to heat bath water.

Regardless of its origin, the end product methane molecule is still very much a pure commodity. But since the production method matters due to the carbon footprint, the pricing of the renewable and conventional methane differs widely. In some Scandinavian countries, subsidies for bio sourced methane is roughly 200% of the conventional natural gas price, showing that the state is valuing sources of methane different depending on the production method. Another country with large scale biogas ambitions have been Germany which has provided so call feed-in-tariffs for biogas, meaning that producers are guaranteed a certain price for their produce by the state. These subsidies has been so lucrative that by 2013 Germany was the worlds largest producer of biogas despite its relatively small agricultural by global comparison.

There's also a trend that companies are willing to pay an additional green premium for biogas to improve their green credentials, and are doing so by buying up green certificates together with their ordinary gas consumption. Since the gas is already subsidised by governments, the green premium is usually pretty small and is thus relatively easy for companies to pass on the additional cost to their customers. Another trend is that municipalities, and even general consumers are buying up the green gas for the sake of lowering their carbon emissions or to support local industries.

Looking at these trend, it seems that several parts of society is starting to value methane differently depending on where it originates from. Going forward there might be the case that a similar trend takes hold for conventional natural gas as well—where for instance gas produced in the EU could entail more value for some patriotic Europeans than imported gas from somewhere else. May it be to ensure energy security, in order to not support foreign non-democratic countries or something else that matters for the consumers. Such a trend cannot be observed so far, and it is perhaps more likely that renewable methane will be the trend that increasingly matter for conscious consumers.

The differentiation of the commodity methane is already a reality, and with ever increasing information density in products and relative ease of passing on additional costs to consumers—it is likely that the decommoditisation trend of energy gases will continue and even broaden going ahead.

In this chapter we have now seen that energy, the epitome of commodities, is in the process of being decommoditised in several ways. The process has arguably just started, and it is currently unclear what the long term consequences may entail. It may be that electric power from renewables will in itself become a commodity due to consumer information fatigue—or the opposite extreme that consumers will pick their energy source specifically both in terms of type, but also regarding the actual production plant unit.

In either case, due to increased data proliferation and information density it is very likely that the decommoditisation trend will continue in some form or another within the energy sector. It may also be that we can actually use the experience and take the learnings from the sector for application in other commodity value chains. This will be explored in Chap. 5.

References

© OECD/IEA (2017) Global EV outlook, IEA Publishing. Licence: www.iea.org/t&c; as modified by Springer. https://www.iea.org/publications/freepublications/publication/GlobalEVOutlook2017.pdf. Accessed 10 Sept 2017

Bloomberg New Energy Finance (2016) Evolution of PV solar panel cost. http://www.bbhub.io/bnef/sites/4/2016/04/BNEF-Summit-Keynote-2016.pdf. Accessed 10 Sept 2017

Intergovernmental panel on climate change (2014a) Main sources of global greenhouse gas emissions. https://www.ipcc.ch/report/ar5/wg3/. Accessed 10 Sept 2017

Intergovernmental panel on climate change (2014b) World energy supply. https://www.ipcc.ch/pdf/assessment-report/ar4/wg3/ar4-wg3-chapter4.pdf. Accessed 10 Sept 2017

Swedish Energy Agency (2017) Energy characteristics. http://www.energimyndigheten.se/fornybart/solenergi/solceller/ & http://www.energimyndigheten.se/fornybart/vindkraft/marknadsstatistik. Accessed 10 Sept 2017

U.S. Energy Information Administration (2016a) Levelized cost of electricity. https://www.eia.gov/outlooks/archive/aeo16/pdf/electricity_generation_2016.pdf. Accessed 10 Sept 2017

U.S. Energy Information Administration (2016b) Carbon dioxide emissions from power generation sources. https://www.eia.gov/tools/faqs/faq.php?id=74&t=11. Accessed 10 Sept 2017

U.S. Energy Information Administration (2017) Capacity factors. https://www.eia.gov/electricity/monthly/epm_table_grapher.php?t=epmt_6_07_b. Accessed 10 Sept 2017

5 Learnings from the Energy Market Decommoditisation Process

The decommoditisation trend in the energy market was not driven by any specific market player nor country, and it took much of the sector off guard even though they could see it on the horizon. In the following chapter we will explore the progression of the trend, how it impacted the market players and try to draw some general conclusions which can be applied across different sectors by creating a decommoditisation framework.

5.1 How to Understand the Process of Decommoditisation in the Energy Sector

It's difficult to pin-point the main reason or starting point for the decommoditisation trend in the energy sector, but there were for sure several catalysts. An important reason was undoubtedly the immense externalities involved in energy production and consumption that was simply not factored into the market.

The energy commodities has arguably the far most extensive negative externalities on the environment and human health, compared to other commodity sectors. It thus became very difficult to conceal this fact in an era of accelerating data proliferation, where public concern and hence political pressure forced the industry to change.

The concern for the negative externalities from energy production had been growing for decades within environmental circles, but was popularised by in the early 2000 century throughout larger population through proponents such as Al Gore which was the major breakthrough for the trend.

Another fact that's striking with the decommoditisation of the energy sector is that it evolved both rather quick, and penetrated very broadly once it had taken hold. Far quicker than most experts within and outside the industry predicted. In fact most forecasters was off by several factors of magnitude when it came to both deployment and cost development. The British regulator DECC's solar PV price forecast in 2008 for 2015 was for instance approx 500% higher than their 2012

F. Andrén-Sandberg, *Commodities Pricing and the Bulk Trap*,
https://doi.org/10.1007/978-3-319-72468-3_5

forecast just 4 years later. Their 2011 forecast was furthermore also abut twice as high for 2015 as their 2012 forecasts. The takeaway here is not that the UK regulator is specifically bad at doing price forecasts for solar PV CAPEX costs. It's rather that almost everyone in the field, from the International Energy Agency (IEA) to several well renowned consultancy firms were of by several factors for magnitude just a few years into the future. The IEA boldly forecasted an increase from 15 GW installed solar PV in 2008 to 410 GW in 2035. In the end of 2016, solar capacity stood at 300 GW with an above 30% increase from end of 2015 with no signs of subduing, and thus ample room to reach 410 GW already before 2020.

> It's very difficult to make forecasts, especially regarding the future

Similar errors were being made across the board on the cost of batteries, wind power deployment, Power-to-gas CAPEX etc. The speed, momentum and broadness of the trend was massively underestimated by most due to the trends own feedback loops—where increased deployment leads to increased learning and price reductions, leading to increased deployment and therefore even further price reduction and so forth.

From a geographical standpoint the epicentre of the decommoditisation process arguably occurred in Europe, which installed the largest amount of solar, wind power and other RES during the early years of renewables—where countries such as Denmark and Germany were clear forerunners. A reason that the trend was catalysed in Europe was probably due to several coincidences such as traditionally strong environmental movement, but also that power markets had largely been liberalised which spurred competition and enabled the market forces to play it's part.

However, as the technology development and information awareness spread, other countries such as the USA, Japan and China was quick to catch up—and as of 2017, most countries in the world from Australia and Brazil to Ethiopia and Russia were undergoing energy transformation in some form or another. Even North Korea has a renewable energy program!

Knowledge Box: North Korean Renewables

Although North Korea is not well known for it's international cooperation efforts, in the area of climate change it has actually taken some measures. The country is facing many of the same environmental challenges as other poor nations, such as soil degradation water management and power shortages. Lately the country has invested in renewables such as solar panels in the cities, and wind energy production in the countryside. Solar panels have become an especially important renewable, since consumers can use them to balance the unreliable national power supply, and the imports of solar panels from China is increasing quickly. The North Korean state media has

(continued)

also reported that it is developing its own solar panel, and its own wind power turbine. Pyongyang even has it's own solar PV powered ferries which offers relief to commuters during the morning rush-hour.

Even thought the renewable energy deployment is North Korea is relatively modest—the interesting takeaway is that not even hermit states are left unaffected by global commodity developments.

5.2 The Incumbent Reaction: Successful and Unsuccessful Approaches

The energy market in 2007 was a very different place than it was in 2017. In Europe which can be seen as the initial epicentre of the disruption, the incumbent energy utilities had been consolidating and were investing quite ambitiously in natural-gas fired power plants. This was seen as the future to lower CO_2 emissions in relation to the older coal-fired units. The energy utilities furthermore had among the largest market caps of firms in Europe, and was very dominating within their fields. Fast forward a decade, and some leading energy utilities in Europe had lost up to 80% of it's market value. The economic crisis in conjunction with massive renewable build-out, which is illustrated in Fig. 5.1, had slashed wholesale power pries to very low levels and forced utilities to mothball brand new plants and make enormous impairment on their capital base. A few of the largest utilities was almost in a state of bankruptcy in the height of the crisis, but most managed by tilting their business to new growth areas, brining in more capital and generally getting rid of loss making assets.

Some was more fortunate than others, but on the whole a lot of utilities suffered greatly and was in a poor position to utilise or drive the renewable decommoditisation trend forward due to their financial difficulties. Some of the energy utilities arguably even fought the trend, as they tried to retain some of the value of their capital base.

The main reaction from most utilities was to invest prudently in the new technology as an additional possible growth leg in their already large conventional generation portfolio. Although many probably did see the new renewables as something that could be a major part of the market down the line, but what took most by surprise was the speed of which the trend progressed. Rather than a curiosity growth leg that could be examined for possible further implementation in 2030, prices for renewables and consumer demand for them progressed at a never before seen pace which made them the primary investment class within energy 2016. This underestimation could also be seen in which manner the renewables was spoken of—namely as 'Alternative energy'.

Toppling the temple can be painful, if you try it from the inside. —Lesson of Samson

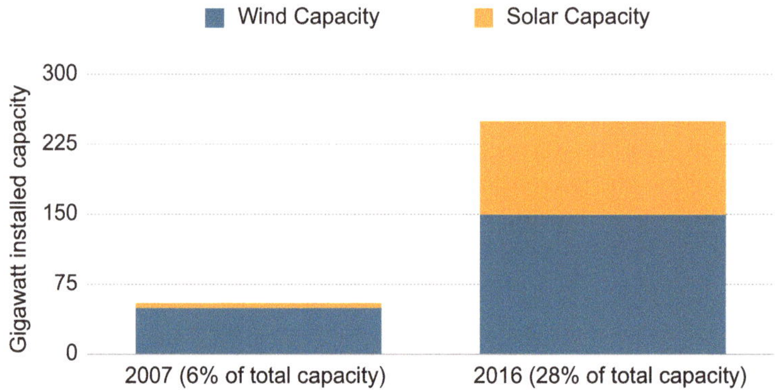

Fig. 5.1 Development of wind and solar energy within The European Union from 2007 to 2016. A build-out pace not predicted by the industry and which has completely transformed the industry landscape. To note that capacity is not the same as energy volume, and that renewable volume share is significantly less than capacity due to lower capacity factors for intermittent renewables. Source: Wind Europe (2016), European power statistics

It might be easy to criticise the utilities who underestimated the pace and momentum of the transition, but the advisor consultancy firms, governments and industry associations that gave them input also got it completely wrong—where many had to revise their price forecast yearly by full multiples, as the market progressed several times quicker than projected.

When the utilities finally realised that a major shift was ongoing, many started to invest in all kinds of consumer oriented products—such a demand response devices and other alternatives. Many of these proved to be failures, as the consumer simply did not demand such services. A possible reason for this may simply have been that the utilities lacked experience of consumer oriented products, whereas they had traditionally focused on commodity products for international markets. Apart from the market failure of some consumer oriented offerings, the plethora of initiatives did bring forward some interesting green sprouts that have yet to prove themselves, but that may show to bear fruit down the the line.

This somewhat lagging reaction opened up possibilities for new market entrants which filled the consumer demand void left open by the incumbents. This was perhaps healthy for the industry as a whole, but the stumbling efforts of the incumbents might also have slowed the decommoditisation as they did not utilise their vast competence in the field to drive the trend forward.

However, it was not a uniform reaction from the industry and many energy companies did for instance invest early in wind power projects—but not at a scale that was in relation to the full potential of the market. Most utilities were betting on CO_2 reductions to come through gradual introduction of renewables, in conjunction with massive build-out of gas-fired power plants—an area where they possessed core competences and were happy to compete within.

Those incumbents who fared better than others were in general those which had more diverse portfolios, both in terms of geographical distribution and technology sub-sectors. Many did for instance have power grids, which has not lost their market value in the trend (yet) and base load power stations which could provide some revenue for the embattled groups.

What the utilities did have though, was plenty of experience within large scale power generation, and lately the wind has started to shift somewhat as the large multinational utilities are once again beginning to gain an upper hand through scale advantage and specialisation (their traditional rationale for existing). Most energy utilities now have decommoditisation as their core agenda where they tailor their energy offerings to their customers—being solar panels to homeowners, electric car charing solutions or huge offshore wind parks for government tenders. Most utilities are now even pushing the trend forward with all their might, trying to convince governments to increase the renewable build-out pace so they can invest even more in renewable energy.

Just as many other industries, the utilities have also started to shift focus away from traditional R&D and putting a larger efforts into innovation and start-up activities such as accelerators and in-house incubation. Some of the innovation activities could be seen as just rebranded R&D, but much has also been focused on more disruptive technologies and new business models that had not been tested before. The incumbents have also started to put greater efforts into partnerships with startups as well as other established industries to be able to compete in the new reality.

In North America, the situation was somewhat different compared to Europe as cheap shale gas had completely flipped the market logic, and many energy utilities were investing in new gas-fired power plants in disfavour of coal. In 2000 approximately 1% of US natural gas production originated from shale formations, in 2010 over 20% of natural gas came from shale, and is expected to increase towards +45% within the coming decades. This development had a significant impact of carbon emissions, which the 2014 per capita emissions being at a 50-year low, but was not driven by environmental concern, but rather pure LCOE economics. However, this development also slowed the introduction of renewables since they became comparatively more expensive vis-a-vis natural gas alternatives.

However, since the cost for renewables were reduced due to deployment in the rest of the world, renewables started to make its steady inroads in the US as well when consumers started to demand them. Utilities have now deployed large scale renewables in states such as Texas, California and many others regions. North America also happens to have some of the most competitive renewable energy resources on the planet—with the 2017 cost of solar PV and wind power in many states being significantly lower than conventional power generation costs. Some American states also introduced carbon reduction targets due to public pressure, further speeding up the process of deployment from the utilities. Consumers were also eager to personally do their part in the energy transformation, which opened up the fields for plenty of new-comers in the business catering for the new consumer palate.

The energy market reaction in many other parts of the globe came bit later than the European experiences. However, utilities across the world have as of 2017 started to, not only catch-up, but also taking a leading role in the transformation in countries such as China, Chile, Brazil, Australia and India—with the cheapest new solar PV contract to date was signed in Dubai 2017 with prices roughly half that of conventional generation costs.

Although traditional energy players has now taken a progressive role in the energy transformation, the initial stumbling of the incumbents made the entrance possible for new players in the industry—and the first part of the transformation was rather driven by new-comers happy to respond to the changing market trends.

5.3 New Value Pools and Specialised Market Entrants

The incumbents arguably slow reaction, made it possible for new companies to emerge within the energy industry. In the US, the most visible examples were of course electric car manufacturer Tesla, First solar, Nest and the likes. However, there were also a plethora of less visible market players catering to the new customer value pools. In Europe the early investments in renewables was made by farmers, municipal energy companies and property owners, rather then large multinational utilities. To cater for this new demand, plenty of non-traditional players popped up to grab a piece of the market, from specialised wind developers and builders, to small scale solar installers and biogas machinery suppliers.

> The future belong to those who sees opportunities before they become obvious

The emergence of supply chains for renewables also initiated a flurry of activities from suppliers. Over time these suppliers and sub-suppliers tended to become evermore consolidated, where the major industrial players such as Siemens and General Electric at later stages even bought up some of the suppliers to further consolidate the sector and reap the efficiency of scale.

Although many of the more established players in the field have grabbed a piece of the market through acquisitions and in house development, the new playing field also created large new industrial giants such as Vestas and Enercon within wind energy and several other firms within the solar panel supply chain. Tesla is probably the best known brand within electric vehicle production, but there are many other large employers within the energy landscape that would not exist if it weren't for the recent trends in the industry.

It's worthwhile considering that the main reason that new market entrants could actually enter the energy fields, what that the incumbents did not see of refrained to act on the new fundamentals such as cheaper batteries which had changed the competitive landscape. Many of the established sport car brands such as Porsche, Lamborghini or Ferrari could very well have produced an electric racer which might had taken the market by storm if they had committed themselves. However, they might have just been a bit too successful in their established market niche to dash out into a new direction, which thus allowed newcomers with no established market to loose, to enter the field. What's perhaps more puzzling is not that so many of the incumbents failed to see the opportunities coming, but rather how slow their reaction has been once the trend crystallised.

There are still many new companies popping up to either supply services to energy companies, or even found niches for themselves competing with the traditional utilities in several categories—selling everything from home installation solar panels, to smart solutions for energy management in homes and offices. It is likely that many of these startups and smaller players will fail, others will find niche markets and grow, some will be gobbled up by established market actors, others will have their business concept copied by incumbents and a few lucky ones might become as large and successful as Tesla or Vestas.

Since the broader decommoditisation of the sector should encompass ever more information density in the products, there are likely to be new niches opening up in the years to come—especially in those areas which caters directly to customer demand and succeeds in transferring the value of some underlying technology such as solar energy or wind in an efficient way. For example enabling direct sale in-between small scale producers of energy, hosting visual comparisons between consumers environmental impact or something completely different but also information related.

One should also remember that the information driven trend has just begun, and the catching up and consolidation that has occurred within the less data driven products in the industry, will probably be replicated within data driven sales and consumer products as well.

5.4 Isolated Trend of the Energy Market: Or a Sign of What to Come Throughout Commodity Driven Value Chains?

The decommoditisation trend that is currently ongoing in the energy sector was made possible due to data proliferation and large visible negative externalities, but also by the option to 'somewhat easy' change into a new path for the industry on an initially small scale. The latter part is important, since changing a huge industry in one blow can be quite difficult, while starting small scale and gain an advantage by improving once position is far more likely to succeed.

Although the energy sector did have all these ingredients, it's not very likely that it should only be an isolated industry trend. Rather, it's quite likely that similar transitions has or will start in other sectors as well. The ones most prone to this disruption are those that have all the right ingredients in place, and is probably where most change will happen the quickest. But over time it is unlikely that any sector will be spared by increased data flows and information proliferation

In regards to ingredients for decommoditisation, there are plenty of commodity sectors that have extensive externalities which could spur the trend. The food industry is an obvious candidate as it's a very areal intensive industry, has a large impact of water supply, soil condition, sea habitat, animal welfare and so on. It is also a sector where consumers continuously demand new products, and where new market players relatively easy can enter to cater for the new palates. Information spread from the sector should also be relatively easy to facilitate, and some countries might even have an incentive to promote the trend.

In contrast, production of industrial solvents might also entail significant externalities, but it would be difficult for alternative players to enter the field and cater for new consumer demand at an initially small scale, since investments in production facilities would probably be very capital intensive and thus hampering to the trend. It would also be difficult for consumers to directly promote the trend, since they seldom buy the industrial solvent commodity directly, but rather support the industry through the purchase of other goods containing a multitude of commodity inputs where the industrial solvent is only one small part.

Another factor that should significantly be able to accelerate the decommoditisation trend is if there is technologies that over time has to potential to actually lower the cost for the utilisation itself. An obvious example would be that of renewable energies with initial high costs compared to alternatives, but where increased deployment has lowered cost to a point where the conventional industry find it increasingly difficult to compete.

This is the key important difference between technologies and commodities. Commodities tend to get more expensive as demand increases, due to limited supply and even exhaustible supply in the long run. If for instance demand for oil increases, more expensive fringe sources needs to be taken into production, thus raising the price for the commodity as a whole since price is governed by marginal price setting.

Technologies on the other hand operates in a very different manner, since they tend to get cheaper the more they are deployed due to the so called learning curve, or learning effect. If the learning factor is say 20% (such as in the case for solar panels), a doubling of output will bring down the cost by 20%. This trend has perhaps been most visible in the semiconductor industry where processing power has increased at en exponential pace, and the price per processing power unit has plummeted.

This difference between technologies and commodities has some profound consequences for decommoditisation. If you introduce 1% of 'new technology solution' (hereafter referred to as NTS) into a commodity driven market, costing twice the amount the conventional commodity. The end consumers of that will pay another 1% of their overall bill for that commodity—and will probably not even notice. However, by creating a market for the NTS technology, there is a high chance that the new solution or technology comes down in costs.

This means that another percent market share of the NTS can be introduced with the overall price hike costing less than percentage, and the consumer should in either case still not really notice the price hike. If the amount starts to approach the 10% limit, then it might be the case that the NTS actually costs the same as the the conventional mix or even lower and thus starts to reduce the overall cost of the commodity system. After this final threshold has been surpassed, it would be relative easy for the NTS to take over the majority of the market share due to inherent competitiveness.

This logic is not only confined to technologies, as cost reduction can also simply be a matter of pure scale. For example services such as organic farm products, initially costing more than conventional produce, would come down in cost when the scale of the production rationalises the logistics of getting the products to market.

For example: A farmer produces 10 tonnes of organic corn, which then needs the be stored in separate containers, and needs its own trucks in order not to blend with the conventional corn filled with pesticides, fungicides and its ilk. It can furthermore not be milled together with the standard 2000 tonnes batch of conventional corn, and thus needs to fetch higher prices due to increase costs for milling independently with smaller quantities. However, if the organic corn production increases to 2000 tonnes, then many of these additional costs will be removed and prices can come down to approach conventional produce or even surpass it.

Commodity markets which have large externalities, low entry barriers for new product offerings, high consumer participation possibilities, and who's market can be transformed by technology and scale will thus have a higher likelihood of being disrupted through decommoditisation.

5.5 Introduction to the Decommoditisation Framework

In order to try to categorise and understand a phenomena, a framework can be a useful lens in which to examine and view the certain pattern that has occurred. The aim of the following decommoditisation framework is to create an overview by breaking down the general characteristics and progression of the trend into more easy to understand formats. The framework draws from the experience in the energy sector, and from the partial decommoditisation that has occurred within other commodity driven sectors as well.

However, one should bear in mind that a framework should not be viewed as an absolute truth, but rather as a model attempting to explain an observed pattern to the best of the current understanding. Just as in physics, a model should be changed if further evidence comes to light, and unlike physics, the framework should be adjusted to the specific conditions in the given market situation to suit the analysis.

Decommoditisation Progression Logic: General Timeline

1. Awareness of externalities confined to special interest groups, but advocacy starts to increase
2. Awareness of externalities spreads to larger sections of market
3. Niche consumers are starting to demand regulation and/or new products that mitigates the observed externalities
4. Market and/or politicians respond with measures to spur new offerings above conventional commodity costs (if offerings could have been done below conventional commodity prices; market logic dictates that they would already have been deployed)

5. New niche market gains foothold—both incumbent and new players enters the field to grab market shares, with outcome highly dependent on the risk aversiveness of the incumbents
6. Niche market growth—increasing scale pushes down price for the new offerings which starts to significantly disrupt conventional commodity market profitability
7. Commodity market logic no longer applies as goods are being priced differently according to the additional information attached such as brand, technology utilised, externality mitigation or marketed storyline
8. (Possible future progression) Commodity fully decommoditised, but with potential for recommoditisation in the future when the new customer requirement standards have been evenly distributed between suppliers and differentiation once again becomes increasingly difficult

Decommoditisation General Logic: Increased Likelihood of Decommoditisation If

1. Visible and concrete externalities are involved in production or distribution of the commodity, which can be mitigated at reasonable costs
2. End consumers with low thresholds can easily choose commodity alternatives
3. High commodity volumes, and large revenue flows through market
4. Alternatives to commodities present, but underdeployed
5. Potential technologies exists which can change market fundamentals if high deployment is achieved
6. Low barriers of entry into the market for new entrants

Decommoditisation Geographical Logic

1. Externality awareness initially confined to geographical area, or within socio-economic group
2. Modest geographical spread in awareness, should be seen as tale-tale sign of acceleration. If 3% of consumers in an area can be affected and are spreading their views, then its fair to assume that 3% in most areas can be affected but have just not yet been disturbed. This entails much larger information surface from where the awareness can further spread to other groups
3. When the progression reaches a point where new product offerings are beginning to transform the market, it is likely that awareness and demand will spread to other regions and larger socioeconomic groups as well
4. Regardless of the origins of the trend and the progression will alter the commodity logic of the market, and is bound to affect and continue throughout the entire commodity market until hitting a roadblock, even fringe markets such as North Korea will be affected if connected to the global commodity market

Decommoditisation Business Considerations

1. New opportunity might to be overlooked due to elevated cost estimates; exponential growth pattern should be anticipated and learning rate factored into forecasts
2. New growth trajectory usually starts at a small scale and might go unnoticed for a while by established industry, but exponential growth could fester into the incumbents marketshare and might rather quickly disrupt the industry as a whole
3. Fringe technologies and offerings should continuously be scouted for and acted upon by established industry to maintain long term viability
4. Incumbents will not be the most likely initial drivers of decommoditisation, as they have vested interests and locked in investments where they already are making profits
5. Pushback from incumbent industry likely if the trend starts to significantly erode profits, but unlikely to have much effect over the long term as information spreads and decommoditisation momentum gathers pace
6. Small scale actors likely to pop-up and cater to new consumer demand patterns if established industry do not retake the initiative and service the new market dynamics
7. Incumbents can await trend progression and consolidate the industry when established to reduce risk, but will loose out on value creation since having to pay full market price for established industry
8. Early mover advantages for both incumbents and newcomers

Decommoditisation National Considerations

1. Countries and regions moving quickly and creates long term decommoditisation certainty are likely to reap higher economic benefits as the development spreads
2. Countries which supports burgeoning industries will generally have higher concentration of the business over the long run, as the home market provides local industry with first movers advantage before the trend becomes an export opportunity
3. Government support should not be needed if sufficient market incentives exists, but might help to accelerate a burgeoning trend and create competitive advantage
4. The countries and regions which identifies and promotes the trend will likely gain an upper hand in the long run, but as picking winners can be difficult—awaiting market signals wise option before creating support mechanisms

Economic Growth Considerations
Decommoditisation should offer higher levels of value added for consumers, and thus higher value for the products. However, this trend might also be recorded as inflation in traditional metrics since the same amount of for instance steel is being produced albeit at higher costs.

The value added in form of data should thus not be recorded as an inflator, just as mobile apps or increased marketing are not recorded as inflation. Rather it should be viewed as the inherent value of the products which is being fundamentally raised and thus economic output increased. E.g. free roaming chickens should not be considered equal to the same numbers of caged chickens, as it is fundamentally different goods even though the meat can measured in the same amounts of kilograms or calories.

Furthermore, if negative externalities are removed which has an overall negative economic impact on the society, the effect of decommoditisation disruption on economic output may be considered positive due to shift in economic activity to more productive sectors.

Knowledge Box: The Difficulty to Assess Economic Growth—and the Negative Effects of Externalities

Economic output and growth is usually measured by the output of an economy, but seldom the actual quality of the production. For instance, a country can increase certain food consumption which entails that the population of the country in question becomes increasingly obese. Businesses will then pop up to cater for this obesity trend, in the form of special diet offerings, exercise books, gastric bypass operations and special medicines which combats the obesity related symptoms. All things equal, this will actually increase economic output as demand increases and production follows. However, one could argue that if the people had just consumed less calories in the first place, all of this economic output would have been unnecessary—and the people supplying them could either be taking some leisure time off or produced some other good for the economy with the same or better overall outcome for society. The question is thus, if the increased economic output from this obesity trend example actually raised economic wellbeing?

In the same manner; increased air pollution due to combustion in cities which causes people to buy face masks and to purchase air purifier machines may actually increase economic growth in itself compared to producing the energy from non-combustion sources, all things equal. But has this increased or decreased economic output?

Does economic output tied to the daily removal of snow from the roads in the Scandinavia, and increased heating requirement during the wintertime imply higher economic wellbeing than in Malta, where they simply don't have the need to remove the snow or heat their houses to the same extent?

The removal of negative externalities may thus not directly be measured in the economy as economic growth. But if it reduces costs in some other part of the economy such as healthcare, then the economy should arguably be considered to perform better since labour and economic activity can be diverted to some other more meaningful task than simply cleaning up pollution or treating health issues stemming from diesel smoke.

We have now learned on how the decommoditisation trend has affected the energy industry, and examined what knowledge that can be drawn from these experiences. We have seen successful and less successful approaches by industry players in the field, and tried to create a rough framework of understanding how the decommoditisation trend took hold and progressed. In the next chapter we will look into other commodity driven industries, to see if similar trends are ongoing within other markets and how it could affect them.

Reference

Wind Europe (2016) European power statistics. https://windeurope.org/wp-content/uploads/files/about-wind/statistics/WindEurope-Annual-Statistics-2016.pdf. Accessed 10 September 2017

Parallel Decommoditisation Trends in Non-energy Markets

6

We have now seen the impact decommoditisation has had on the energy sector which has experienced substantial disruption, partly due to the large externalities involved in energy production and consumption. But what about other commodity markets—are there any signs of decommoditisation and disruption there? In this chapter we will look into several unrelated commodity fields to examine if decommoditisation has started to affect these sectors as well, and whether the drivers might be similar.

6.1 The Blood Diamonds of the 1990s: When Origin Is Starting to Matter

An early example of a commodity market that has been partially differentiated is that of diamonds. Diamond is chemical substance which is very difficult to tell how it was produced, where a diamond from Indonesia will look more or less identical to a diamond mined in Botswana or Russia for the end consumers.

Diamonds is essentially just carbon atoms that has bounded together in a certain way in response to pressure and heat. So for the end customer using it as jewellery or for industrial cutting, this would imply that the origin of the diamond has very little relevance, and that the main thing determining the value of the diamonds is supply and demand. Thus, diamonds can be considered a commodity.

However, during the 1980s and 1990s, there was an increased realisation that the diamonds used for wedding rings and other jewellery was sometimes bought from dubious sources, where the people who had been mining them might have been exploited and treated unfairly. Even worse was that the trade in some of these diamonds had in itself become a major revenue steam for warlords and criminals, actually fuelling the conflicts by it's extraction, and thus the concept of "blood diamonds" was born.

Since most people did not want the externality impact of their wedding rings to be fuelling conflicts and harm people in war-torn areas, there was a push from

F. Andrén-Sandberg, *Commodities Pricing and the Bulk Trap*,
https://doi.org/10.1007/978-3-319-72468-3_6

consumers demanding that their diamonds should only be sourced from ethical sources. This push was prior to the large scale penetration of the internet, so most of the awareness spread though news, magazines, movie documentaries and of course word of mouth.

In African countries from Angola to the Ivory coast and the Congo's, blood diamonds were funding terrible civil wars and insurgencies. The organisation 'Global witness' was one of the first major organisations to show the link between blood diamonds and several of these conflicts in a report released 1998. This report was later followed by an UN enquiry and resolutions trying to stem the harmful trade. To enforce the resolutions the so called Kimberly process was formed. This set out to make trade in blood diamonds more difficult by punishing both countries and traders of the products if not in compliance with the standards, and by effectively banning end traders of the products if found guilty of conflict diamond trade. The system also gained credibility among end consumers, who now demanded that their jewellery should be certified by the system.

Although not perfect, and while some corruption have been identified, the process has undoubtedly made it a lot more difficult for dubious miners to sell their diamonds onto international markets. This have in several cases stemmed the income streams which fuelled insurgencies, causing them to calm down significantly while at the same time raising revenues for responsible governments in compliance with the process. Rough estimates states that approximately 20% of the diamonds traded in the 1980s where of conflict origin, compared to just under 1% in 2004. A significant decrease that have had very much to do with changing consumer preference and the institutional change within the trade.

Another development in the diamond industry that has nothing to do with conflict is the introduction of synthetic manufactured diamonds, i.e. diamonds not mined from the ground. Since diamonds is just carbon atoms put together in a certain way, they can literally be made out of a lump of wood, charcoal or the graphite in a pencil. The raw material for making diamonds is thus very cheap and abundant for the small amounts of diamonds needed. The difficulty of making diamonds is rather due to the process; where large amount of pressure and heat is required, or an advanced chemical process needs to be initiated. Very small diamond crystals (nano crystals) can even be made by detonating explosives containing carbon in enclosed chambers, where the sudden increase in temperature and heat causes the carbon atoms to fuse (perhaps out of fear!) into a diamond structure.

Since diamonds is one of the hardest material known to man, manufactured diamonds are today utilised for a range of different applications; such as polishing agents, cutting blades and other industrial needs. However, synthetic diamonds have yet to penetrate the gemstone market, where manufactured diamonds is estimated to only constitute 1–2% of all diamond gemstones sold. Synthetic diamonds usually also fetch a lower price than regular diamonds by about 30%. This even though they are more 'perfect' from a crystal-structure perspective than mined diamonds, and do not have the conflict or sustainability issues that naturally formed diamonds may have hanging over them.

So if looking at these trends, diamonds has actually gone from being a commodity for end consumers to a differentiated product where origins and production processes of the material actually matters quite a lot. Since the end product is more or less indistinguishable, the differentiating factor is the attached commodity information, where certification schemes and credible institutions play a major role. This differentiation and decommoditisation of the commodity has arguably been beneficial, with less negative externalities being caused in the production of the mineral, and the planet is arguably better off as a whole.

The question is if this trend will or could be replicated for other commodities as well. Could oil be certified to avoid stemming from troubled areas that fuels conflicts—such as in the 2016 Syria/Iraq ISIS situation, where oil was a major part of the terror sect's revenue stream. Is it quite probable that similar decommoditisation processes could happen for other minerals such as gold or rare earths metals, where origin awareness and increased information density in the products starts to play a larger role. We will continue to explore this in the coming sections.

6.2 Free Roaming Hens, Organic Cucumbers and Non-GMO Maize: Are We Witnessing the Beginning of the End for Commoditised Food Products?

The food industry is currently experiencing a significant rebound in consumer awareness of its practises. Just a hundred years ago in today's developed world, production of food was a very visible enterprise which most people conducted for themselves. The farming and handling of most vegetable and grain products was usually done very close to home, and the welfare of the hens, goats and cows consumed were very visible since they were living right next to the consumers.

This changed radically in the 1900s, where at the turn of the century more than half of the population in the current OECD countries were involved in food production—at the turn of the twenty-first century, as a contrast, this number has dropped to around 2–5%. This shift was made possible by significant improvements in crop yields and increase scale of animal production.

This agricultural revolution involved the introduction of large scale mechanised farming practices as well as factory production of animals, which lowered prices and increased efficiency of production in terms of yields per capita. This meant that large part of the population could be fed by the few, which enabled the lion part of the labour force to contribute in other areas of the economy and develop extensive manufacturing and services industries.

Knowledge Box: Modern Poultry Farming

Around 70% of the worlds bird production is 'intensively' farmed. Egg laying hens starts to produce eggs at an age of 16–20 weeks, and is slaughtered after one year of egg production as their productivity declines. Battery hens have less then a A4 paper per chicken of living space, and indoor non-battery hens are usually packed together 10–20 individuals per square meter. None of these chickens or hens will ever smell the open air or see daylight during their lifetime. Battery cages was banned in the European union in 2012, but still staple practice in many areas of the world.

Broiler chickens for meat production are usually slaughtered after 5–9 weeks of age, and their growth rate can be as much as three times its own weight in their first living week. The birds are packed into houses amounting to 20,000 chickens per house, and the floor usually rises during contentious production as the animal litter becomes the new floor, and there needs to be plenty of ventilation in order for the ammonia and heat not to suffocate the animals.

The production also cause severe stresses on the birds physiology, and cannibalism are not unusual due to the packed conditions. The unnatural quick growth rates also cause deformities of the chickens, where many have problems walking since their legs become underdeveloped and cannot sustain the fast increase in body weight—essentially becoming crippled and unable to escape local dirt and heat conditions which makes easy prey for cannibalising peers. Many also develop heart and lung problems which can cause early death. To sustain the toxic conditions, constant antibiotics are often administered in order for the birds not to succumb to diseases.

However, chicken and hens do not have to live like this, as the conditions is actually just caused by the active choices made by industry and consumers buying the products. Free range chickens or certified alternatives can be chosen instead and usually have daily access to outdoor areas, with a lot more space to move around with subsequent less or no need for antibiotics. They are usually also allowed to breed more slowly, causing less negative physiological effects on the chickens. The drawback of these more well managed animals is the higher cost of producing eggs and chicken meat, but several consumers seem to believe it's worth the extra cost as demand for these products are increasing in several markets.

However, the flip side of this industrial transformation was that while efficiency increased, the long term sustainability the farming actually decreased due to negative externalities. These externalities included; over-fertilisation of crops leaking into the water system causing over-nitration and dead zones in the sea, significant losses of bio-diversity, mass deaths of natural pollinators, soil depletion,

Fig. 6.1 European indoor free range egg laying hens. With the packed together to increase production and decrease costs; causing severe stresses on animal welfare. Copyright: Author

and many other kinds of other negative externalities causing harm to both nature and long term economic prosperity.

Animal welfare was another aspect that arguably did not improve with industrial farming. For instance, in modern farming 16 hens can be packed together per square meter causing cannibalism, some pigs are unable turn around due to the small size of their boxes, cows being stuck in the same 6 square meters for their whole life, upbringing of cage hens, broiler chickens, and so forth. The crowded conditions for some chickens can be seen in Fig. 6.1. Another aspect is that the quality of food in many cases are also reduced; with less nutrition, less taste, less texture and in general just less appetising food for consumption.

Because of these developments a counter trend has been growing, where organic and ecological food, free range chickens and others more empathetic options such as the the example in Fig. 6.2 have experienced a significant upswing during the last decade. An interesting part of this trend is that it has actually been active for quite a while, but confined to sub-populations of the consumer market which has cared a lot about these issues for decades. But the trend did arguably not really take off until the onset of the internet and social media, where the easy sharing of information, pictures and maybe most importantly; 'videos', could be distributed to raise awareness of animal welfare and food related issues.

Social media is ripe with videos and articles being shared regarding food production and consumption, and it is not anymore easy to hide unethical farming practices such as low labour standards or rainforest clearcutting to make way for crops. Concerned consumers who are looking for alternative consumption sources are also being catered for by companies which offers produce with less externalities, such as organic crops or free range outdoor hens. These companies can now also share their more sustainable practices through social media and thus being able to fetch the higher prices necessary to sustain their production costs, as indicated by the higher prices for organic corn illustrated in Fig. 6.3.

Fig. 6.2 Outdoor raised free range pigs in Europe. Which fetches higher value for some consumers; knowing their food items having lived a decent life. Copyright: Author

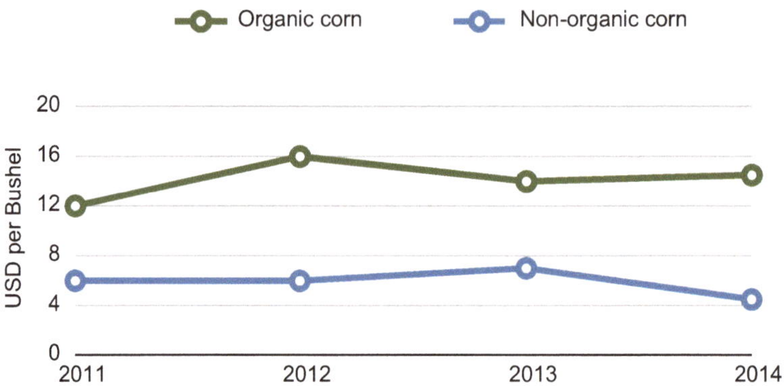

Fig. 6.3 Yearly average prices for organic and non-organic corn in the US. Showing not only higher prices for organic, but also more resilience in the 2014 conventional commodity price fall, indicating differentiation of the commodity. Source: U.S. Department of Agriculture (2014), Organic Prices

Knowledge Box: UN Food and Agriculture Organisation (2017a), Why Is Organic Food More Expensive Than Conventional Food (Cut-out)

"Prices of organic foods include not only the cost of the food production itself, but also a range of other factors that are not captured in the price of conventional food, such as:

- Environmental enhancement and protection (and avoidance of future expenses to mitigate pollution). For example, higher prices of organic

(continued)

cash crops compensate for low financial returns of rotational periods which are necessary to build soil fertility
- Higher standards for animal welfare
- Avoidance of health risks to farmers due to inappropriate handling of pesticides (and avoidance of future medical expenses)
- Rural development by generating additional farm employment and assuring a fair and sufficient income to producers

Secondary consumers (e.g. bread producers) of these products have also been taking notice, identifying the need to source their input materials from more sustainable sources in order to not loose market share, or just to increase profit margins of the produced products.

This secondary market for farm produce is particularly important, since commodity input costs in the end products usually comprise a very small share of their total production costs. This has the implication that even if the primary inputs such as flower for bakeries or sugar for sodas is twice as high if buying organic than the conventional variety, the total increase of the product price may be as low as 5%, if the commodity inputs only makes up 5% of the total initial production cost. Since marketing, logistics and packaging costs often widely eclipses the input commodities costs in food products, the producers can often increase prices more than the additional costs to the consumer, thus creating an even better margin than for the conventional product.

The higher prices for organic end customer products is illustrated in Fig. 6.4, with organic on average around 35–45% more expensive than regular products. Main reason for higher prices are:

- Limited supply compared to demand (increased producer margins)
- Greater diversity of enterprises and thus more difficult to achieve economies of scale
- Cost increases due to post harvest segregation of produce and relatively inefficient logistics attributed to smaller volumes

Implying that larger volumes could decrease end customer prices significantly due to mitigation of reason 1, 2 and 3.

Since consumers are willing to pay more for higher end produce and companies are turning better margins on them—there are substantial incentives to grow the market and for companies to promote the products. Research on the subject actually shows that even though organic farms have lower productivity per acre of land for

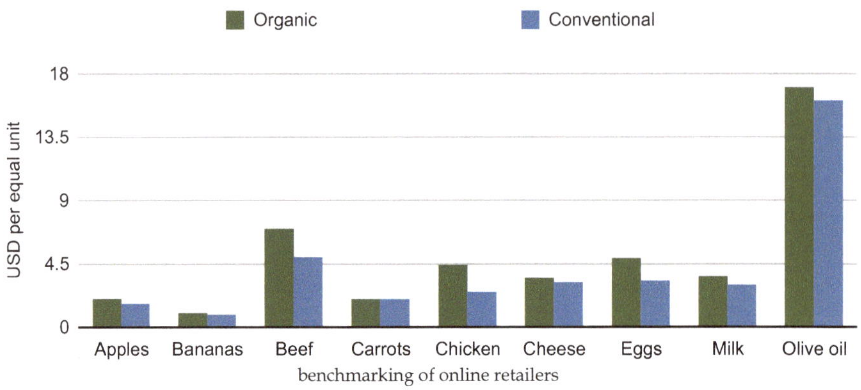

Fig. 6.4 Organic versus conventional products at end customer supermarket. Source: Author (2014), benchmarking of online retailers

their crops, they are on average more profitable than their non-organic alternatives, suggesting that they are able to fetch better returns than for conventional produce. Average crop yields for organic and non-organic produce can be seen in Fig. 6.5.

A recent example of this is can be found in the Nordics; where consumers who are unsatisfied with the supermarkets general offerings are taking the distribution into their own hands, by creating so called REKO rings (REKO being a finish abbreviation for ≈ "Good, Nice and Fair"). The concept is that consumers come together with local food producers through Facebook groups, where they order their food of the certain quality they demand. The farmers then distribute this food at certain locations and days from the trunk compartment of their cars or similar. The value added for the consumers is that they can receive better information on how their food have been produced, where it have been produced, and by whom. When conducting these affairs, the customers spends quite a lot of time acquiring the goods, and also pays more money purchasing them compared to conventional produce since the goods are usually of higher cost origin through organic or ecological practices. The farmers are allegedly also making a higher margin on the produce, since able to differentiate the products from the bulk volumes which they are used to sell through.

As an example; there was in 2017 a Swedish tomato producer who instead of using pesticide, used natural insects and animals to keep the pests in check. The farmer also heated his glasshouses with local firewood instead of fossil fuels, all in all giving the customer a sense of supporting sustainability when making the tomato purchase. If these tomatoes would have been sold in bulk at the supermarket instead, without any of the 'cool insect pest control information' attached, the customer would not have been able to experience the value added and thus treat the tomato as just another commodity. The farmer would thus suffer from the higher production costs without being able to increase prices above other producers

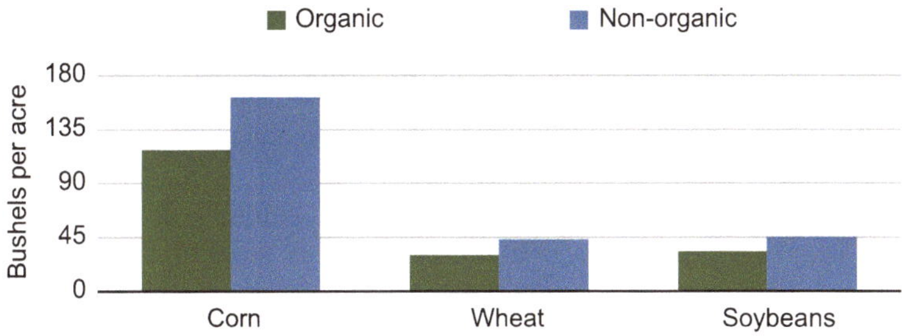

Fig. 6.5 US crop yields for organic and non-organic crops per unit area. Showing lower yields for organic produce per land area. However, lower yields does not imply inability to feed the global population, it just means that less land area must be diverted to animal production or animal feed production if not wanting to increase agriculture land use. Furthermore, less externalities and soil erosion may actually entail higher production yields per acre for organic over longer time perspectives. Source: U.S. Department of Agriculture (2017), Crop production

tomatoes, and the arguable good work of supplying value added would simply been lost if general prices where insufficient to support the business.

Knowledge Box: Permanent Crops, Next to Come After Organic Produce and the Future of Agriculture?

A recent trend has been sweeping the avant-garde sustainability food sector has been the push for permaculture. Permaculture in essence means constant agriculture, and has been attached to various practices pushed by the 'sustainable' movement. A main theme is though that instead of planting perennial crops, which needs to be replanted every year such as wheat or potatoes, food producing crops that remain year after year are promoted, which means less soil degradation and CO_2 leakage into to atmosphere among other things. There are plenty of grasses that might be suitable for this, which can produce flower without the need of being replanted after harvest, much similar to a regular lawn which does not need to be replanted after cutting. Other farmers use trees with nuts which retains the same properties. The problem has so far been crop yields per unit area which is lower than traditional varieties, but further refined crops could very well increase these. Another form of agriculture is so called hugelkultur (German), where wood is buried in the ground which creates soil and sequesters carbon—while at the same time storing water for plants like a sponge with less need for external watering systems. Another aspect of permaculture has been gardening in layers, where there are treetops in the sky, bushes as mid vegetation, herbs on the ground and even mushrooms in the ground to have a large variety of fauna to harvest. This can also be combined with general animal husbandry gracing below.

Main arguments of permaculture are:

(continued)

- Less tillage, would imply reduced CO_2 into the atmosphere as agriculture amounts for large part of CO_2 emissions. Bare earth furthermore increases solar absorption, worsening the effects of global warming.
- Less soil erosion, which is a natural process but speeded up by human activities 10-40 times the natural base line. If nothing is done, there could be a real lack of cultivatable soil available in the future. Permaculture would reduce this significantly, as trees supplies food annually from the same trunk, and grasses with deep roots simply regrows after harvest, without need for plowing.
- Biodiversity, offers constant habitats for ecosystem and animals, and large variety of different outputs could be co-hosted in the system. Examples of this is raising animals under the trees, mushrooms from tree symbiosis and low level vegetation which supplies nutrients to the soil.
- Economically sensible to have grain grasses that are perennial, which then just grow year after year just as a common lawn instead of costly replanting. These usually also have deeper roots, which might increase drought resistance.

The REKO ring trend started in Finland 2013 (drawing inspiration from France) and spread to Sweden in 2016. As of 2017, roughly 1% of the total Swedish population were involved in REKO rings—up from 0% the year before. The trend furthermore continue to experience rapid growth, and can be seen as yet another example where consumers are willing to pay a premium and bear quite some inconveniences for certain goods when the produce is attached to specific value adding information about those goods.

Similar trends like the Nordic REKO rings can also be found globally, with the common differentiator of food commodities usually revolve around ethical practices, eco-sustainability, locally produced or all of the above. However, although growing rapidly, if looking at the organic crop share of total production as a proxy for the decommoditisation trend in the food industry, the trend is then still very much in its infancy.

The total share of organic crop acreage in the US is still only around 1% of total produce, although the sector did experience an astonishing 240% growth increase from 2002 to 2011 (1.3 to 3.1 million acres). However, in countries such as Sweden and Denmark, organic food produce as of 2017 already amounts to approximately 9% of the total sold volumes. The market is also experiencing sustained double digits yearly growth rates, with up to +45% year-on-year growth within certain segments. Organic produce from Swedish agriculture furthermore amounted to 17% of total grown crop area in 2016, with government targets to almost double the organic arable land towards 2030. The public sector in Sweden have been also been mandated to spur the growth by setting targets such as +60% mandated organic food purchases in 2030 (from 32% of total consumption 2016). Targets such as these increases investment certainty for producers, who can rely on a solid base segment while expanding to the wider end-consumer market.

If such growth rates can be sustained and spread to other countries it would not take long for the trend to cause quite an upheaval for the conventional food commodity market within the coming decades. The organic farming growth in conjunction with higher animal welfare standards for meat products have by itself create increased employment in the food industry, higher value creation for the economy and more sustainable food sourcing for society. Although it is far from certain that this trend will take over the entire food market in the foreseeable future, there are substantial economic rational and plenty of incentives supporting the transition in the longer term, and might be ill advised betting against this trend taking further hold of the market.

6.3 Seafood: Sustainably Sourced from Certified Fisheries

Another food variety that has historically been commoditised is different sorts of seafoods. Compared to fishing just a hundred years ago, the variety and fish species available for most consumers have declined drastically, and fish are usually sourced globally rather than locally to satisfy the societies massive demand for high protein food. Seafood is also the largest agricultural export commodity from developing countries, outpacing coffee as the second largest agricultural commodity export by a factor of two, and is thus a very important export revenue for many developing nations.

The insatiable demand for seafood have also introduced the concept of industrial fishing, where some ships are more or less floating factories that catch fish and turn it into packaged end customer products for the market while on the sea. The problem with this practice is that the ocean unfortunately does not contain an endless supply of fish to catch. However, due to insufficient regulatory bodies governing the global stocks and since the fishing industry have been mostly concerned with short term profitability, the incentives have alas been skewed for fishermen to catch as much fish as cheap as possible before the other fishermen does.

Knowledge Box: Floating Fish Factories

The introduction of floating factories ships and fish processing vessels were made possible by modern freezing technology, where fresh produce are less necessary and ships can go further away from home ports since being able to store their catch for a longer time.

These modern ships can catch enormous quantities of fish, and may have fishing nets covering over 2 km in circumference. These nets have the ability to catch entire shoals, leaving no fish left in the area to repopulate the stock.

(continued)

Another issue is that the large nets also catches a lot of by-catch, such as Dolphins, deep water corals, fungi and other marine critters. Some deep water fungi have life spans of 2000 years, and do not reproduce at very quick rates.

The first examples of similar ships (without freezing ability) was actually whalers, who went along way from home port to pursue their catch, and thus needed to process the whales carcass into the more profitable whale oil to save bulk space onboard the vessel. These whalers decimated some whale stocks close to extinction before the international community stepped in to control the industry.

This practice have led to overfishing, with globally fishing stocks being depleted, ecosystems collapsed and fish at some instances even hunted close to extinction. It's not very difficult to understand why; if 50 tunas are born every year in a certain area from a population of 100 tunas, it's simply impossible to catch more than 50 tunas a year without the stock being depleted in the long run—as the birth rates would be reduced every year from fifty, to forty, to ten, to nil.

Silly as it may sound, this is actually what has happened in many areas in the world, and what's even more worrisome is that even after the fishing of certain species stop, the stocks may not even be able to fully recover within reasonable timespans since the ecosystem balance have been fundamentally disturbed. When for instance cod populations declines under a certain threshold, other species that used to be their pray may increase in numbers and starts to eat the young cods hatchlings with greater intensity. Meaning that the cod population are unable to recover and control the population of the now abundant species lower down in the food chain, which can have an adverse effect of populations further down the food chain and so forth. It's difficult to see how the balance can be restored, unless migrating populations or different species arrive to fill the ecosystem gap, which might take some time.

Knowledge Box: Newfoundland Cod Fishery Collapse

Outside Newfoundland, Canada in the early 90s, cod fishing had been conducted for around 500 years, and was a vital part of the regional economy. But after seeing fishing stocks falling to 1% of its historical levels, the Canadian federal government in 1992 after a heated debate, finally declared a moratorium on cod fishing. The reasons for the collapse was the intensive and ever more efficient fishing practices, with larger ships and more sophisticated technologies in order to track schools of fish at deeper depths and being able to catch them for longer periods. The catches was well over the replacement rate, and subsequently the stocks simply collapsed.

(continued)

The consequences were severe, with over 35,000 fishermen and fish processing plant workers becoming unemployed in over 400 costal communities, and despite the banning of cod fishing the population has as of 2017 not yet bounced back to healthy levels. The costal communities have adopted, however, where some have shifted to crab and shrimp fishing which increased their populations as a consequence of the decline in its cod predators. The sea ecosystem has unfortunately not been as fast to adapt as their human counterparts.

There are several fish species which are now close to complete collapse, but there's a problem with shifted baselines where society usually only compare current catches to the last 5 years historically very low volumes, and not to populations volumes 30 or 100 years ago. Eels are good example where stocks are now only 1% (!) of what it was a hundred years ago, but where fisheries usually compare current eel stocks to the last few years when deciding what volumes to catch. This becomes very evident when looking at the historical eel stocks in Fig. 6.6, where arguable the most rational course of action would be to stop the fishing until the stocks recover to acceptable levels. The primary reason this practice have been able to continue, is that no entity actually owns the international fishing resource. This entails the business rational that if your own ship don't catch the fish, then somebody else's ship probably will—twisting the fishing incentives into cricket swarm logic.

The overfishing practice is also very wasteful in a purely economical sense, where restoring stocks in the EU alone would increase catches of approximately 3.5 million tonnes a year—worth around EUR 3bn per year. Since reducing fishery income in the short run by significantly less than this, there would be as solid business case for any entity if they were just in charge the whole system.

From 1970, seafood production and catches have grown at an annual pace of 8% per year, and catches increased tenfold from 1950 as seen in Fig. 6.7. One third of all fishing stocks globally have already collapsed, which is defined as a decline to less than 10% of their original stock, and are thus having a hard time to bounce back even if the intensive fishing would stop. At current fishing rates, all stocks(!) are expected to collapse within a few decades according to the UN. As fishing stocks collapses, trawlers search out new ecosystems such as deep sea creatures to maintain volumes. The problem with these 'new stocks' is that they usually have much lower replenishment rates, and thus recover much slower from intensive fishing than for instance fish such as herring or mackerel. In the long run though the practice will come to a roads ends, as there will simply not be any more additional populations to deplete.

Today the most commonly fished seafood species are: shrimp, salmon, tuna and cod—and subsequently, all of these stocks are well under their historical levels. The over-exploitation of the fishing resource in conjunction with devastating fishing

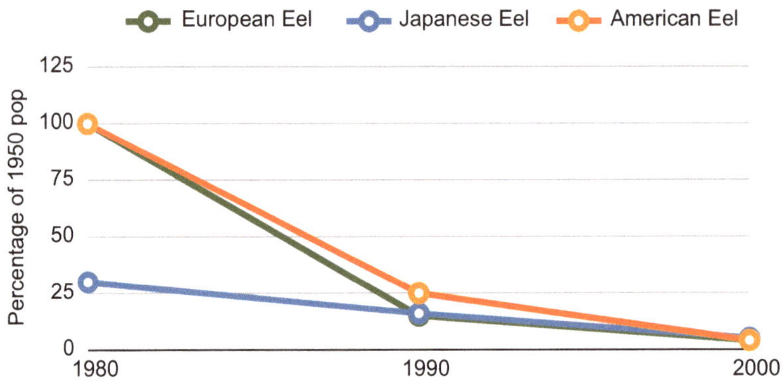

Fig. 6.6 Percentage of total 1950 juvenile population of eels. Having declined towards a few percentages of their historical stocks. With over 85% of the total global fishing stocks over exploited or depleted, similar trends can be seen across populations of cod, tuna, whales and many other sea living populations, especially for animals higher up the food chain. Source: Food and Agriculture Organisation of the United Nations (2017b), Fishing trends and stock data

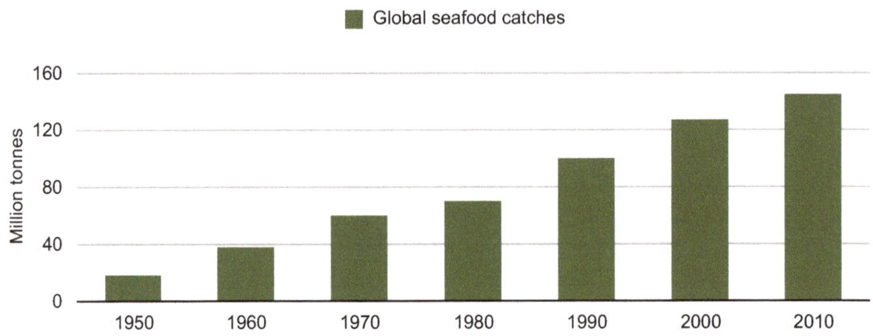

Fig. 6.7 Global fish catches as production volumes. Figures including aquaculture which often receives feed from conventional fishing**.** Catches has experienced an almost tenfold increase from 1950s levels, which should bring some food for thought. Source: Food and Agriculture Organisation of the United Nations (2017c), Global Capture Production

methods such as bottom trawling or reef fishing using poison and explosives, have had a profound impact on the marine environment.

Another problem is that large trawlers from developed nations obtain fishing rights from poorer nations to fish outside their coast lines. This have in many cases had a devastating effect on local sustainable fishery industries by depleting local fish populations, which can easily be seen in the statistics, with costal fisheries having declined by over 50% in the last 30 years. This trend can especially be observed in west Africa where some coastal communities relying of fishing has been devastated and families being torn apart when breadwinners are being forced to relocate into the cities to find jobs and sustain their basic needs.

However, this bleak reality with plenty of negative externalities caused by modern fishing have caused many consumers to simply stop eating fish, or at least demand that their fish are being sourced from places which are being replenished and fished in a responsible manner.

This phenomena is quite new though. Up until recently, very few people cared if their frozen fish was coming out of the Atlantic or the Pacific, if the fish had been farmed or caught in the sea, or whether if it was fished by bottom trawling or by hook. However, consumers are becoming ever more aware of the problems with modern fishing, and many consumers are simply not willing anymore to partake in the long term depletion of our common ocean resources. In response to this consumer trend—there have sprung up plenty of companies around the globe which are managing their fishing resource in a long term sustainable manner, with good fishing practices that do not cause excessive harm to the environment where it operates.

A problem, though, is that since most fish is not sourced from the local environment, consumers have mainly been limited to base their purchases on container labelling, which states where they fish were sourced from and contain reassurances from unknown actors that these fisheries are being well managed. However, it might be difficult for consumers to probe into the validity of the claims and to make a proper judgement if a hook caught Alaskan pollock is in fact sustainable as the vendor claims it to be, or whether the practices are in fact bit fishy. It can become quite a daunting task for consumers to keep track on what species from which areas, being caught by those fishing methods that are deemed long term sustainable.

A solution to this problem has been the development of certification bodies such as MSC branding of fisheries. MSC stands for Marine Stewardship Council, and is a global non profit organisation who's goal is to ensure sustainability of fishing with thriving and productive oceans in the long run. The organisation evaluate fishing methods and fish stocks to determine whether it's sustainable or not, and then grants their logo to the products if it lives up to their standards. The branding has become grantor of quality and makes it a lot easier for consumers to pick the products that are in line with their priorities.

A complementary branding ASC, which stands for Aquaculture Stewardship Council, is filling a similar function but for farmed fishing practices. The idea to farm fish instead of hunting them in the ocean may sound like a brilliant idea at first, but the problem historically has been that the farming in many cases had a substantial negative impact on the marine environment close to where it operates. Another problem have been that the fodder for the fish in many cases have just been ordinary fish caught in the ocean, which has been grind up and turned into pellets. With the entailing issue that it takes a lot more caught fish to 'create' the farmed fish, and would in theory thus be better to just eat the caught fish right away from a calorie perspective.

Additional issues with fish farming is that they can be energy intensive, sometimes very polluting and may need to use antibiotics and other anti-parasitic drugs in order for the fish to grow and flourish. Furthermore in many tropical regions, mangrove coast line have been cleared to make room for shrimp farms—which part from destroying the mangroves that is the breeding ground for many ocean dwelling creatures, it also creates polluting run-offs that has adverse effects on the local marine environment.

Fortunately, there are responsible farmed fishing practices available as well, with well managed stocks mainly fed on vegan diets and which are causing a lot less pollution and adverse side effects. The fisheries who does utilise responsible and sustainable farming practices can get certified through ASC and be able to sell their products using their brand. This makes it a lot easier for customers to pick the products that fits their ethical values without knowing specifics regarding the exact production source.

The key issue of these types of certification brands is of course credibility. There are several certifications schemes currently in use for fisheries, but since many of them have been created by the fishing industry itself, consumers might, however, find it difficult judging the credibility of the labels. Consumers are generally wary of green washing (insincere labelling or story lining for marketing purposes), and the need for certifying agencies to be transparent and unaffiliated from industry has often been the key for attaining credibility and gain consumer trust.

Although sustainable fisheries often have higher short term costs than conventional methods, the increased costs can usually be offset by an increased consumer willingness to pay, and producers may even receive higher margins than conventional commoditised fisheries. This applies especially to those who in addition to certification have focused on transparent information regarding their business practices, which they successfully communicate to consumers. Part from the additional revenues in the short run, the long term financial value of sustainability should not be underestimated. This since it is arguably also from a business standpoint quite unwise to catch 100 tunas a day if the populations is only growing by 50 tunas a day; hence depleting the producing stock and not being able to catch as much as otherwise possible in the long run.

In 2015 less than 1% of the ocean were so called protected area, where fishing is banned or severely limited in conjunction with other protective measures. However, countries around the world have started to take note and by 2020 the international community has agreed to increase ocean reserves to 10% of total ocean surface area. These reserves are very important as it gives breathing space to pressured species, and it has been estimated that designating just 4% of the worlds oceans to protected status could protect up to 80% of all marine mammals species. There have also been good examples of river cleanups and restoration of habitats, where fish have been brought back to thrive in previously ‘dead rivers’. It is for instance now possible to find salmon in the English Thames, which was almost unthinkable just a few decades ago. So there is hope, and if fish can be bought back to thrive in former dead zones, coral reefs and oceans should also be possible to bring back into better health if fishing practices just change onto a more sustainable path.

Plenty of consumers are beginning to understand this logic, and are now voting with their wallets. Fishermen who can prove their sustainable credentials with differentiated products and perhaps even local connections, and are able pass on the value added information to their customers will generally be better off than conventional commoditised fish producers. The market for sustainable seafood is experiencing double digit growth in several countries, and is expected to catch an ever larger share of the overall market in the coming decades. It hence seems like the seafood market is once again becoming decommoditised and reclaiming the diversified nature that fish and fishing previously has been and actually is. While this trend progresses—consumers, the oceans and the industry as a whole will probably all be better off in the long run.

6.4 Wood and Biomass: Externalities Revealed and Consumers Are Taking Note

Lumber and general wood are another commodity category that is currently experiencing a differentiation trend. Forests make up around 30% of earths total land area but has regressed significantly over the last decades—continuing a trend that has been going on, albeit more slowly, for several millennia. In the past, forests was cleared to make room for agriculture and pastures, as well as to supply fuel for cooking and heating. Wood was also a primary building material for houses, furniture and ships, and due to all the convenient uses; the forest resources came under severe pressure in many areas across of the world. Europe in particular have seen an enormous change in the topography of the continent, where most of what is now agricultural land in both Germany and Britain used to be covered with thick forests. Due to increased population growth in conjunction with forest clearing, the forest area per capita in 2016 was just 7% of what it was in early 1900 as seen in Fig. 6.8—and about 1% to what it was just a few hundred years ago.

Due to the high demand of forest products and land clearing practices, forests have been in decline and is still declining most areas of the globe. The only land areas that are currently experience marginal net forest growth are located in Europe and North America, but are growing quite slowly. The problem with the general decline is that forests and their trees are quite handy as they among other things; creates oxygen, binds carbon dioxide, hinders soil erosion, creates soil, manufactures nutrients for plants and animals, creates local climates and habitats of animals and plant species, and are the source of many medicines we currently use today. The rainforests of South America even produces its own rain, and South America without its forests would arguably look very different.

Forest are thus quite important, and cutting them down have a severe impact on the long term wellbeing of animals and humans alike both locally and across the globe.

The collateral damage of habitat destruction can also cause a severe impact to local ecosystems and decline in species populations as seen in the Fig. 6.9 for Orangutang's as a notable example. The main reason for this specific decline is habitat loss due to forest clear cutting, where 80% of the Indonesian old rainforest were logged from 1980 to 2005, and many of the previous forest areas in Borneo have been converted to palm oil plantations, which severely limits orangutang habitat and their survival prospects.

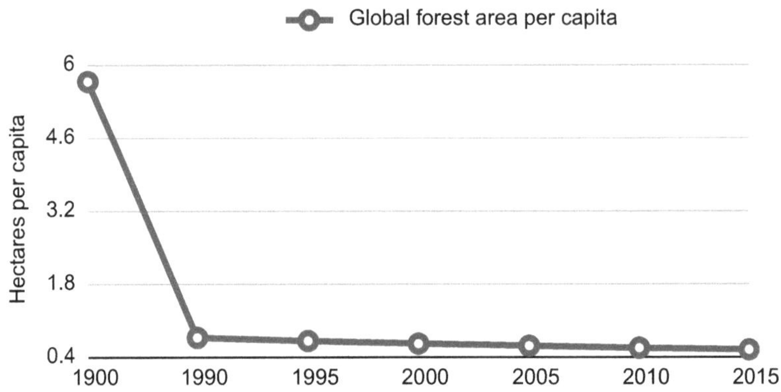

Fig. 6.8 Global forest stock per capita. Which has declined dramatically and steadily in the last century and expected to become less than 0.4 Hectares per capita in 2050—roughly 15 fold lower than the ratio of the early 1900. Source: Food and Agriculture Organisation of the United Nations (2017d), FAO Forestry Statistics

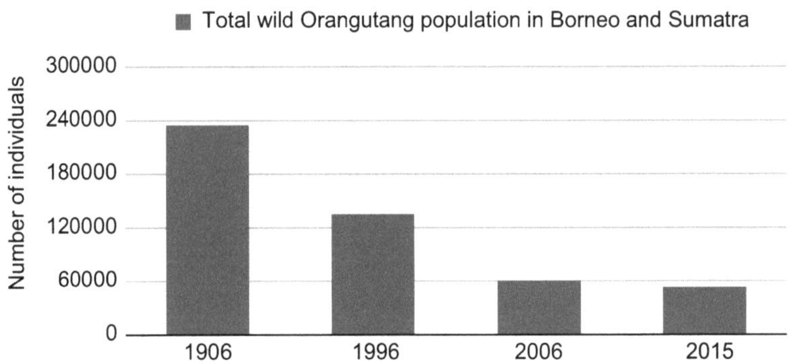

Fig. 6.9 Total wild orangutang populations living in Sumatra and Borneo. Which have been severely diminished, with the Sumatran orangutang population only numbering at around 7500 animals. Some young orangoutangs are also taken as pets, and sold on the black market in South East Asia. Source: © 2017 WWF (panda.org). Some rights reserved. Orangutang populations living in Sumatra and Borneo

However, the unsustainable practices throughout several rainforest areas, where forests have been made way for crop plantations or pastures, has lately gained a lot of public attention. Palm oil plantations which crowd out animal habitats have particularly been criticised for its practices, and some consumers have taken action by demanding palm oil free products or sustainable palm oil in their products.

In terms of lumber, there are several initiatives to ensure sustainable sourcing of forest products such as the FSC certification (Forest Stewardship Council), which certifies both end consumer products such as paper and wooden board, as well as wood fuel used for industrial combustion. Several countries now demands that the energy producers who uses biomass in their power and heat generation, can prove that their sourcing is sustainable in order to get green subsidies for their output. As of 2015, the amount of planted wood which is certified had 50 folded since the early 1990s and continues to experience rapid demand growth.

However, in this commodity market is not so much that the certified sourced biomass fetch much higher prices than non-certified biomass at the end consumers. It has rather become a binary issue where some consumers, companies and nations simply refuse to purchase the products unless it's certified or otherwise proved sustainable. A less binary and more classical decommoditisation part of this trend is that wood origin is also beginning to matter more for some consumers, where certain biomass origins fetches higher value than other. French produced lumber may for instance to some consumers be worth more than imported Russian lumber, and can thus actually be able to fetch a higher price. The same differentiation is ongoing for small scale sales of wood pellets and other biomass energy related products, where brand, source and quality is starting to have greater importance in the eyes of customers.

The circumstances regarding forest related production have started matter for many end consumers, and it seems like the differentiation of the commodity is gaining momentum. Several niche application can already be considered products, where brand and attached information is staring to play a greater role. Given the ever increasing information density of products and services, the pivotal ecological importance forest, and the subsequent enormous impact forest management has on our planet's well being—there are reasons to believe this decommoditisation trend will continue to intensify even further.

6.5 The Evolving Plastics Landscape: The Onset of Decommoditisation for Ordinary Input Materials

The plastics commodity have very much transformed how we live in today's world—from cheap packaging and toys, to durable car parts and water piping. Plastics has provided a cheap, reliable and 'easy-to-mould' material for the manufacturing sector. Plastics are made from single molecules called monomers, which are made to react and form long chains of the repeating molecule to form what's called polymers. These chains, which are usually made from a carbon molecule backbone, are situated next to other similar chains to form a sort of micro spaghetti structure, which at a macro level becomes the plastic. The word plastics actually means 'flexible', and can be manufactured to have tailor made properties in terms of melting points, chemical resistance, rigidity, strength and so forth, just depending on the length of the molecule chains and type of monomer used in the process.

However, the problem with plastics has mainly been what also makes it so useful—it does not easily break down in nature. Plastics in nature can take anything from 1 to 500 years to break down, depending on the type and thickness of the material. If not properly disposed, plastics may end up in forests or blown into the sea—causing problems by aggregating in large garbage hoops due to water

Table 6.1 The degradation time of products in the oceans vary depending on composition

Type of product	Degradation time in ocean
Paper towel	2–4 weeks
Newspaper	6 weeks
Cardboard	2 months
Photodegradable beverage holder	6 months
Plywood	1–3 years
Plastic bag	1–20 years
Tin can	50 years
Foamed plastic cup	50 years
Aluminium can	200 years
Plastic bottle	450 years
Fishing line	600 years

Source: Swedish country administration association (2017, Degradation Times (Länsstyrelserna)

currents. This have a major impact on marine life since the plastics are being eaten by birds and fish alike, causing congestion and even poisoning of the animals. All albatrosses examined in the Midway Island in 2015 where found to have plastics in their digestive systems, with many of the young chicks dying as a consequence from it. Fish and other ocean dwelling creatures suffers the same problems, with the additional risk of breathing the partially decomposed micro plastics, which can interfere with bodily functions such as gils and suffocate the animals. Plastics in the oceans are thus creating havoc on ecosystems and many citizens now demands that steps are to be taken to ensure long term viability of several species.

The degradation time of plastics in the ocean can be seen in Table 6.1. The numbers can be disputed to whether its 50 years for a plastic bottle or 450 years, depending on when the plastic is considered degraded—where a plastic bottle might not have a plastic bottle form in 100 years, but may have been broken down to micro plastic pieces which can linger around the ocean for a long time further and cause problem for marine wildlife. What is less disputable is that the buildup of many of these materials in the oceans is a problem, and that even 100 years degradation time is still very long.

Another problem is that most plastics today are made from fossil feedstocks such as oil or natural gas, so even when properly incinerated there's still a net increase of CO_2 emissions to the atmosphere. A solution is to landfill the plastics, but is a very wasteful use of the resource and not sustainable in long term. Recycling of plastics such as PET bottles are taking place and collected on a large scale, but since the composition of most mixed plastics is so diverse, it can be difficult to turn it into new material of the same quality—entailing that much of the source separated plastic materials being collected today is still heading for incineration.

However, consumers have taken note of this problems are demanding solutions. One of these solutions has been the introduction of bio-plastics, where the fossil based feedstock is substituted for renewable biomass.

These bioplastics can be put into two different categories, where the first category is bio-plastics that behave exactly like ordinary plastics and does not have any major characteristics that sticks out in terms of function. The main thing that differs from conventional plastics is that the material is sourced from renewable feedstock and thus won't run out, and that the combustion of the material should not result in major net CO_2 emissions. But there's still the problem that much of these plastics can end up in the ocean, and continue to wreak havoc on the marine environment.

The other category which solves this is the biodegradable plastics. These plastics inherit many of the characteristics of ordinary plastics and will not emit net CO_2 when combusted, but also have the additional benefit that they will break down and degrade quicker if released into the natural environment. The usual drawback of biodegradable plastics is that they are less durable than ordinary plastics, since one of their key design criteria is to in fact break down, and therefore cannot replace ordinary plastics for all applications.

Knowledge Box: Plastics or Polymers in Nature

Plastics is a synthetic polymer, meaning a long chain of the same molecule that forms a mesh with other polymers like a molecular spaghetti—creating a structure that has high durability and can take a lot of mechanical strain.

Nature have also utilised polymers to solve many of its on own challenges, such as cellulose in wood or starch in our food. The main difference between synthetic polymers (plastics) and natural polymers such as wood or straw is that the latter has co-evolved in nature with organisms that can break them down, such as bacteria and fungus or in the case of starch—humans. This in contrast to for instance PVC plastic which few natural organisms were used to deal with just 100 years ago, and have thus had insufficient time evolving to utilise the carbon compound as food.

Most plastics are today made from fossil feedstock, but could also be made from renewable forest biomass feedstock. Fossil fuels are after all just old partially decomposed biomass material, so the producers might as well skip the million years intermediary and go directly to the carbon source.

However, the duality of the two bio-plastics is also a key sales argument. Since many of the plastic products made today are made out of the convenience of the cheap and easily mouldable material, and not because the products actually needs to maintain its durability for the next 100 years. Children toys, toothbrushes, car parts and other goods that are made to last, should perhaps be made from non-degradable bioplastics since they need to endure harsh and wet conditions over many years. Although degradable bioplastics could definitely be used for short term packaging in areas such as food wrappings, shopping bags, styrofoam replacement and other items that are non-essential parts of the product consumers buy, but rather a container meant to be thrown away. It would thus be beneficial for the plastic to just degrade in case it would end up in places where it was not intended.

There have been some controversy and debate regarding the sustainability of bio-plastics as well, since the feedstock in many cases are being grown using farm machinery that uses petrol, excess use of fertiliser, etc. In some instances meaning only 20–30% lower CO_2 emissions compared to oil based alternatives. This may be true is some cases but the feedstock itself will still be renewable and the farm machinery does not have to run on petrol, the crops could also very well be grown using sustainable practices—a value chain cleanup potential that cannot be replicated by the bioplastics fossil cousins. Furthermore the biodegradability of the bioplastics is another sales argument which are harder to achieve with fossil based system, although it's in theory possible.

The positive impact of the introduction of these products should not be underestimated, and even though they only constitute a tiny part of the total plastics market, the product category are experiencing very high growth rates in several markets. As consumers are beginning to change their purchasing habits, companies have also realised that they may be able to increase the profit margins on their products. Although bioplastics cost more than ordinary plastics, is usually only comprise a very tiny part of the overall value chain cost for the product. For instance, plastics bottles costs around €2–3.5 cents more per bottle if using renewable plastics, where the actual feedstock constitute roughly half of that cost. Even a 50% increase of bioplastics prices would therefore not make a significant dent in the end product price, and the €3 cents in extra cost can easily be offset by a €5 cents increase on the end product (approx. 2–4% increase of total ordinary price product in the case of a soda) if including labelling that clearly states the bottle's green credentials. This means that producers can actually increase their margins on their products, while consumers can make an easy choice between green and conventional—potentially also showing their surroundings that they care. Cost of bio-plastics differs widely, but are usually less than twice the cost of ordinary plastics.

Governments have also been taking note of this trend, and some cities such as New York, Washington DC and Miami has gone so far as to ban or partly ban styrofoam plastics. This since styrofoam does not degrade very quickly and had thus become a major waste problem in these cities, as styrofoam has been used for everything from packaging to disposable food containers and doggy bags. Other regions has followed suit, and the EU have for instance included bioplastics in their waste legislation strategy as a key pillar in the transformation into a circular- and bio-economy. In the private sector, large consumer beverage companies such as Pepsi and Coca cola have introduced renewable packaging, including a 100% bioplastic PET bottle in 2015. Heinz and other large food producers seems to be following similar strategies in pursuit of the trend.

Although growing rapidly, bioplastics currently only make up about 1% of the total plastic production volumes. However, it is forecasted to experience double digit growth in coming years, and since the business and environmental rationale speaks if favour for its implementation, it seems like a trend that is bound to continue its expansion in the coming decades. A logical next steps due to the varying sustainability of the different bioplastics, could also be that consumers start to differentiate between sources of bioplastics as well, even further decommoditising plastics.

6.6 Fair Trade Branding: Why Coffee Might Never Become Just Coffee Again, and Why the Norm Is Back!

Ever increasing globalisation and the subsequent production off-shoring has meant that the spectrum of labour standards in place for producing products and commodities have widen dramatically compared to when most goods were produced domestically. Unfortunately the divergence has seldom been for the better, and the working standards for those who produce goods and commodities are in most cases significantly lower in the countries where production have been off-shored to.

Many products such a clothing and toys are manufactured in large scale factories, where labourers may work very long hours for relatively low pay. Even more concerning is that many of the workers are exposed to unsafe working environments, with accidents and exposure to toxins being part of the deal. Modern day slavery is also present across several value chains, and one should bear in mind that only around 20% of present today slavery can be attributed to prostitution and sex trade, with the rest being exploited for production of regular goods and services.

There are many cases where poor practices have been exposed, some of them through extreme tragedies such as in the 2013 Bangladeshi garment factory collapse. In this case a garment factory overcrowded with workers which far exceeded the buildings design limits, invoked a structural collapse which caused 1130 people to die. With the arguably main cause—the intent to produce goods as cheap as possible for rich consumers in the developed world (approx. 2500 people where saved from the rubble). When the appalling working conditions became apparent, consumers unleashed their anger against the retailers sourcing from the factory, forcing many them to clean up their supply chains and temporarily had a negative impact on their business reputation.

Concern over working practices and exploitation of workers in third world countries and increased media attention has created pressure on companies to clean up their act. It has thus been vital for primary manufacturers improve their operations in order to retain their retail company customers who in turn is concerned over their brand perception. However, in the eye of the consumers, it has been relatively difficult to know which companies who are just 'green-washing' (fair-trade washing) their practices and who is really doing something substantial. A solution to this have been the introduction of fair-trade branding schemes, where non-partial certification bodies makes sure that the products is being produced in an ethical and sustainable manner.

Knowledge Box: Global Consumers Willing to Pay More for Sustainable and Social Responsible Services and Products

A Nielsen global survey from 2014 interviewing 30,000 consumers in over 50 countries showed that 55% of global online consumers were willing to pay more for goods produced and distributed by companies committed to positive social and environmental impact. The analysis also showed that the trend was on the rise and over 50% as likely in Asia and Latin America (~65%) compared to Europe and North America (~40%).

Furthermore, products branded as sustainable on packaging experienced twice as high year-on-year sales increases on average compared to non-branded products, and five times higher for products that promoted sustainability throughout marketing platforms.

Perhaps even more important for the future was that Millennials (age 21–34) represented 51% of the respondents who was actively searching out sustainable labelling. Furthermore, they were three times more in favour of sustainability actions than Generation X (age 35–49) and actually 12 more likely than baby boomers (age 50–64) to be in favour of sustainability actions.

An updated report from 2015 also showed that positive respondents in Generation Z (aged 15–20) where up from 55% in 2014 to 72% in 2015! If this trend maintains momentum, then this segment will represent an ever increasing share of the overall market in the future—with substantial impact on sales of these products.

"Brands that establish a reputation for environmental stewardship among today's youngest consumers have an opportunity to not only grow market share but build loyalty among the power-spending Millennials of tomorrow, too," says Grace Farraj, SVP, Public Development & Sustainability, Nielsen."

Source: Nielsen (2014/2015), Global Survey on Corporate Social Responsibility

Much of the fair-trade certification have so far been focused commodity farming; making sure that for instance coffee growers have decent working conditions and fair pay for their labour. Coffee is one of the most traded commodities globally in economic terms, although the farmers who grows it in many cases earn less than $2 per day. This in conjunction with harsh working conditions and long hours has made it difficult for the farmers to escape the poverty trap. The idea of fair trade is to increase the pay and improve working conditions for workers across the value chain, by selling the end product at a higher price than the non-certified conventional commodity, and thus releasing more funds for pressured workers. However, in order to do this, the value of promoting better working conditions must be incorporated into the goods and transferred to the end customers. This is being done through the visibility of the certification brand, but also through transparent

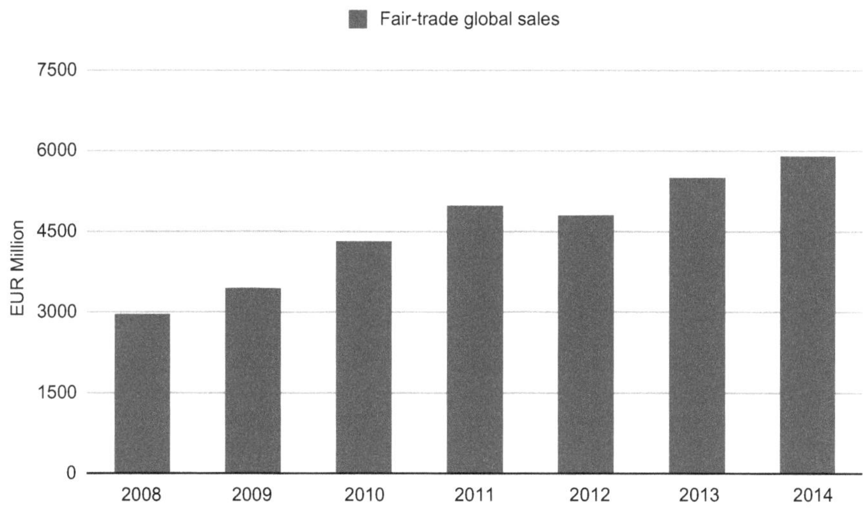

Fig. 6.10 Fair-trade international global sales volumes. Which on average have experienced double digit growth in the last 7 years, although with high varieties across market—with 2013–2014 growth rates in Sweden and Germany topping 37% and 27% year on year respectively. Source: Fair trade international (2017), FAO Annual reports

information channels and campaigns to raise consumer awareness regarding the worth of the certification. This increased information density entails additional costs to the end price of the products, but customers seem to consider the goods to be well worth the extra costs.

Fair trade is actually not a new trend, and have been promoted by various religions groups as early as the 1950s, but it is not since recently that the trend as taken off and created a major impact in the market. The Fair-trade international global sales volumes shown in Fig. 6.10, can been seen as a proxy for this recent broader increase, almost doubling yearly sold volumes in under 6 years. There are now even fair trade cities that promote themselves as socially responsible. The partial reason for the recent 'take off' is arguably due to the penetration of internet, social media and the subsequent increased information density attached to products regardless if the producer wishes it to be so or not. Another reason is that there is arguably better absolute margins to be made on this business model as illustrated in Fig. 6.11, where the same margin leads to overall higher profit for the involved parties due to higher turnover of the goods. Moreover, there are usually substantial room to increase margin levels due to niche markets, further increasing incitements for business to enter and push the field.

Unethical production practice videos can now be spread virally, causing uproar among certain consumer segments but also increased peer pressure on other consumers segments to vote with their wallet on the issue. This is causing differentiation in the commodity market for several primary commodities such as coffee and wheat, where origins and production methods is once again starting to matter

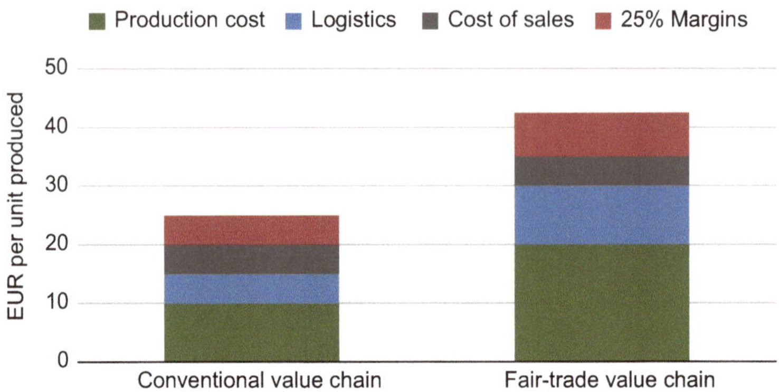

Fig. 6.11 Conceptional comparison of conventional and fair-trade value chain. Applying same margins yields increased profitability of fair-trade value chain due to the higher overall revenues—there is thus a substantial incentive to promote fair trade value chains from a business perspective as well as an ethical one

for the end consumers. Although many fair trade certificates are mostly attributed to finished products, they are also applied for commodities since the whole value chain is usually taken into account.

Fair trade issues are arguably also more visible and easier to relate to compared to for instance global-warming issues—as its relatively easy to feel compassion for labourers in overcrowded workshops or poor farmers tilling their land with outdated equipment.

Due to this empathetic response and increased transparency of value chains, it is likely that this trend will continue in one form or another, whether it being through strong brands or through certification bodies. There is a lot of people who stands to benefit if it does continue, and should also be possible for companies to reap higher returns due to the increased overall turnover. Hard to beat incentives which probably will further accelerate the fair trade trend.

We have now seen the impact decommoditisation has had on several non-energy markets, with a spectra stemming from the diamond industry and agriculture to fishery and plastics. Furthermore, there are few reasons to believe that the trend would be isolated to the markets which are already impacted. Its rather likely that increased information proliferation and data flow will push decommoditisation into other sectors as well. In Chap. 7 we will examine a few candidates which could be ripe for such information disruption.

References

Fair trade international (2017) FAO Annual reports. https://www.fairtrade.net/about-fairtrade/annual-reports.html. Accessed 10 September 2017

Food and Agriculture Organisation of the United Nations (2017a) Why is organic food more expensive than conventional food. http://www.fao.org/organicag/oa-faq/oa-faq5/en/. Accessed 10 September 2017

Food and Agriculture Organisation of the United Nations (2017b) Fishing trends and stock data. http://www.fao.org/fisheries/en/. Accessed 10 September 2017

Food and Agriculture Organisation of the United Nations (2017c) Global Capture Production. http://www.fao.org/fishery/statistics/global-capture-production/query/en. Accessed 10 September 2017

Food and Agriculture Organisation of the United Nations (2017d) FAO Forestry Statistics. http://www.fao.org/forestry/46203/en/. Accessed 10 September 2017

Nielsen (2015) Global Survey on Corporate Social Responsibility. http://www.nielsen.com/us/en/press-room/2014/global-consumers-are-willing-to-put-their-money-where-their-heart-is.html & http://www.nielsen.com/us/en/insights/news/2015/green-generation-millennials-say-sustainability-is-a-shopping-priority.html. Accessed 10 September 2017

Swedish Country administration (2017) Degradation times. http://www.lansstyrelsen.se/skane/SiteCollectionDocuments/Sv/nyheter/2013/Plast_Nedbrytningstider.pdf. Accessed 10 September 2017

U.S. Department of Agriculture (2014) Organic Prices. https://www.ers.usda.gov/data-products/organic-prices/. Accessed 10 September 2017

U.S. Department of Agriculture (2017) Crop production. https://www.usda.gov/topics/farming/crop-production. Accessed 10 September 2017

World Wide Fund for Nature, WWF some rights reserved (2017) Orangutang populations living in Sumatra and Borneo. http://wwf.panda.org/what_we_do/endangered_species/great_apes/orangutans/. Accessed 10 September 2017

Commodities Next in Line for Decommoditisation

7

We have now seen several commodity markets being disrupted by decommoditisation. But what about the other markets seemingly not yet disturbed, what signs should we pay attention to for us to determine if decommoditisation pressure is building up. In this chapter we will examine just this, and which likely candidates that might be next in line for disruption.

7.1 Why Some Commodities Have Yet to Be Disturbed, and the Early Indications That They Might Be Disrupted

In the previous chapters we've seen examples of several commodities in the process of being decommoditised, but the question is which commodities could be next. This is of course very difficult to predict, but so far the furthest disrupted industries has been the ones which are consumer oriented—and we have yet to see fair-trade or sustainable sulphuric acid being significantly fronted towards the market. It is not unfeasible that commodities such industrial solvents will be decommoditized in the future as well. However, it is far more likely that high volumes commodities which is being sold directly to the end customer, or as a major component in what they buy, becomes decommoditised before that.

Another crucial point is there there needs to be significant spectrum of externalities involved in the production of the commodity (positive or negative ones), which is not already factored in by the market. If for instance all electricity were just being produced by the same process yielding similar environmental emissions and having the same impact on society, then there would be much less room for disruption in the industry. That is, unless the consumer can live without the commodity by replacing it with other goods or services.

This last point is important since commodities should also be seen as nothing more than a utility—a good that satisfies a certain need for the customer. Due to this, one should perhaps not only look for differentiation of the commodity itself, but also which other goods that could replace it. If for instance aluminium

F. Andrén-Sandberg, *Commodities Pricing and the Bulk Trap*,
https://doi.org/10.1007/978-3-319-72468-3_7

production would be deemed too much to stomach for consumers, and the aluminium value chain is just too difficult or too costly to clean up—then aluminium could perhaps be substituted by carbon fibre or steel instead for many its applications, if these are considered less disturbing.

A good example of this is oil, in the case of the utility transport. A commodity which is currently being disrupted not by new oil, but mainly by new sorts of transport fuels and services.

In order for decommoditisation to take place, there must also be significant possibilities for increasing diversity in the value chain of the commodity, and possibilities to mitigate or reenforce the negative or positive externalities. Otherwise it would be difficult for producers to differentiate their products, and for having a possible revenue spectrum which is a prerequisite for decommoditisation. It also helps if it's not too costly to mitigate the negative externalities, as the barrier for consumers is then lessened.

The last major factor is that there needs to be externalities involved that consumers actually care about, and an initial sign of decommoditisation would be increased public awareness.

If a certain commodity utilisation is destroying the planets biosphere, are responsible for significant public health problems or are on the verge of causing mass extinction of species on a similar magnitude to the end of the dinosaurs (fossil fuels), then there is a increased likelihood of the public taking note. On the other hand, if the commodity production mainly causes local holes in the ground, which are later turned into pretty lakes which the public can fish within (some stone quarries—oversimplified)—the public response will be more likely limited to NIMBY (Not In My Back Yard) discussions and less likely to have a severe decommoditisation impact on the end customer sales.

To sum up, commodities are be more likely to face disruption if:

- Close to customer end purchase, with high level of consumer control
- Value chain has a spectrum of externalities not factored into the market
- Having possible substitutes to the commodity utility
- Less complicated and not too expensive to clean up value chain
- Has significant externalities, which consumers might care about if known

The questions is then, which candidate commodities do have some of these characteristics and might be ripe for coming decommoditisation?

7.2 Oil: The Most Geopolitical and Economically Important Commodity in the Decommoditisation Crosshairs

An obvious candidate for future decommoditisation would be fossil oil. Since the commodity is relatively easy to store and transport, has a high energy density and is widely available across the globe; oil has dominated the world's energy supply since the second world war. It is also widely used as feedstock material for everything from fertilisers and pharmaceuticals to asphalt and plastics. This flexibility has made oil one of the largest traded commodities in terms of both volume and turnover, and a lot of it is actually consumed (after some refining) directly by end consumers across the entire world for transportation purposes.

The problem with oil is of course its fossil nature, entailing net releases of CO_2 into the atmosphere when combusted, adding to global warming and ocean acidification.

Part from basic CO_2 emissions, the spectrum of externalities from oil production is also quite broad, where Canadian tar sands arguably have a larger environmental impact per litre of petrol than for instance a litre of Saudi produced petrol. On the other hand, the Canadians may find it more appealing to buy domestically produced oil and support local industry, than to shore up the accounts of a non-democratic foreign country. Similarly, patriotic Americans may find it more appetising to consume Texas produced oil and write it on a bumper sticker, than to use Russian or Venezuelan oil. Furthermore, plenty of companies would probably proudly state which fuel their lorries run on, in order to cater to their customers pallets.

Since its a quite diverse commodity with both environmental and geopolitical externalities and thus plenty of room of differentiation, oil and petroleum products actually inhabits many characteristics which would make it ripe for decommoditisation.

Due to the large spectra of sources and composition—crude oil should perhaps even be seen more as a common denominator of several types of extractable hydrocarbons in liquid form, instead of a fixed substance commodity. It varies a lot by appearance depending on its composition and from where it is sourced; from black to brown, to yellowish or even reddish. It is usually found in conjunction with natural gas, which is partly a byproduct of the organic elements breaking down in the oxygen free environment underground.

Oil can in addition to conventional oil fields also be extracted from oil sands, where the oil is stuck in a semi solid state packed in-between the grains of sand. Oil sand actually stems from conventional oil climbing towards the earth surface, and then partially decomposes in exposure to a more oxygen rich environment. Some of the heavier elements are still intact however, and usually possible to recover through clearing the vegetation, unearthing large volumes of top soil, and essentially open pit mine the oil out of the ground. Another method is to extract the tar sand oil volumes by injecting steam into the ground, and then just pump the oil back up instead. The oil is then treated in refineries and converted to byproducts and fuels to be burned for energy; such as heating oil, gasoline, jet fuel and diesel.

Oil types such as these are usually called unconventional oil since they are not extracted through conventional processes. They can be found across the globe, but are especially abundant in Canada and Venezuela. The unconventional oil reserves in just these two countries are roughly twice as large as the global conventional reserves of oil in terms of energy. The drawback of unconventional oils though, is that they require a lot of energy input in order to extract it from the ground. The subsequent CO_2 emissions and environmental impact is thus arguably greater than for conventional oil fields under normal circumstances.

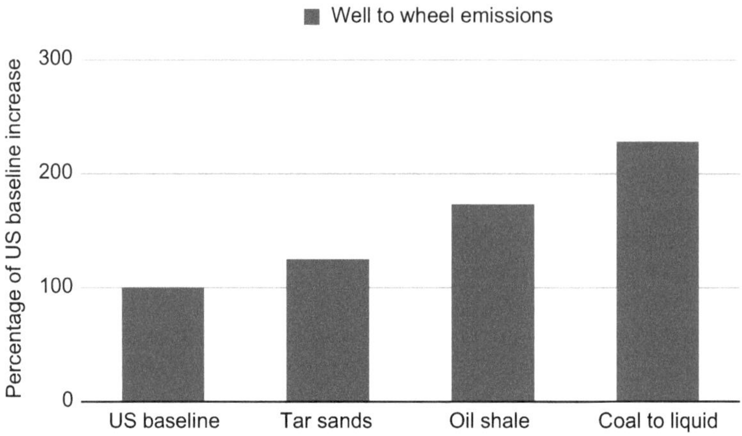

Fig. 7.1 Well-to-wheel carbon emissions from different fuel sources. Source: Included with permission from the Natural Resources Defence Council, NRDC (2010), Lifecycle Emissions of Tar Sands & GHG Emission Factors for High Carbon Intensity Crude Oils

This can be seen in Fig. 7.1, where a litre of gasoline may cause twice the amount of CO_2 emissions compared to the US baseline depending on how it was produced. Coal to liquid as shown in this example is seldom used, but might be feasible in the future if oil prices are elevated high enough and cost of carbon emissions remains low. Worth to remember is also that CO_2 emissions is just one of the environmental impacts of oil production. Tar sand and shale oil for instance uses significant amounts of water, and tar sands commonly uses large areas of land turning it into moonlike landscapes

However, since the energy potential and dividends if extracted at the right cost levels are tremendous, these sources will continue to be mined if fetching the same price on the commodity market as other oil alternatives.

However, another aspect of oil which will eventually force the market to find an alternative is that since it's non-renewable, the market will simply run out. Many oil experts have for a long time warned about peak oil, when oil production will hit its historical maximum to then start its slow decline. According to the theory, this would of have a large impact on the global economy since increasing oil consumption have been fuelling the prosperity of consumers in many developing countries as they gets richer. The decline of oil production would lead to a sharp increase in prices and the entailing global consumer battle for the last precious drops, which would hamper economic development in many areas in the world who could not afford it. However, the scenario would be economic Disneyland for countries with large untapped oil reserves.

If looking at current conventional supply data and current consumption rates, one would conclude that the global oil reserves would last for about another 38 years (from 2017 baseline). If adding unconventional supplies such as tar sands, the number rises towards another 120 years. Factoring in current increase in global oil consumption would lower the figure with a few decades depending on the forecast. Although interesting numbers in itself, this way of counting totally misses the main point of peak oil.

Even if the market would have enough oil to last another century, the peak in production would still have a severe impact on the economy, since the market will exhaust the cheapest and most accessible sources first, and then move on to more expensive supplies (with probable worse environmental profile). These supplies would cost a lot more to extract, and the margins for oil production even with increased commodity prices would thus decline for many producers. Prices for the commodity would then rise until it becomes economically profitable to extract from even trickier sources, which would stabilise the price for the moment until that source in exhausted, then moving on to more expensive sources and so on.

So the question is perhaps not when we will reach peak oil, but rather when we will reach peak cheap oil. The International energy agency stated in 2010 that peak oil has already occurred in 2006. As predicted, in the following years as prices rose in the commodity boom, new sources of oil where coming online at a higher pace such as shale oil in the US and tar sands in Canada. Horizontal drilling and enhanced recovery techniques were also deployed throughout the world to increase output from existing sources, as the increased prices made it economically rational. Global oil production per region is shown in Fig. 7.2

Knowledge Box: Peak Oil?

Peak oil? Both the European and US extraction of conventional oil seem to already have peaked and similar trends has been seen in several other markets. But as prices rises, new sources of oil may become viable and cause production to increase once again. However, the issue may not be peak oil supply but rather peak oil demand—as demand for more expensive oil fade in favour for other more competitive forms of energy sources and carriers such as electric cars and solar or wind produced electricity. The dynamics of these two categories are very unsimilar; as the fossil resources going towards depletion the costs will rise as demand growth outstrips supply, while technologies such as solar and batteries tends to decrease in cost over time due to the learning curve which lowers prices as deployment increases.

The most probably scenario is fortunately that once oil prices exceeds a certain limit, alternatives energy sources and carriers such as electric vehicles and biofuels would become increasingly competitive, thus setting a de-facto price limit of how high oil prices could actually become before demand weakens. Some analysts have even begun to discuss the prospect of peak oil demand, where renewable alternatives will simply squeeze out oil production well before the recoverable volumes have been extracted. The effect of this being that costs of renewable

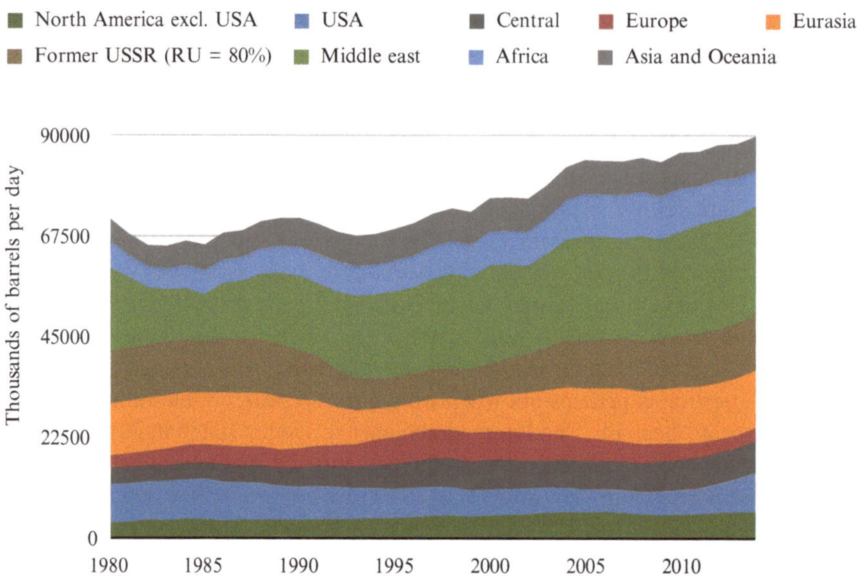

Fig. 7.2 Global oil production per region. Source: US Energy Information Administration (2017), Petroleum and other liquids data & analysis

alternatives will set an effective ceiling on possible oil prices, and thus production volumes if remaining a commodity.

Another more worrying scenario would be that the alternative energy sources may not be renewables but rather conversion of other fossil fuels sources such as coal into gasoline. This would in theory be possible, but entail increased negative environmental impacts compared to conventional oil.

In this post peak cheap-oil world the decommoditisation of oil is even more likely, since the spectrum of production methods and their impact would vary a great deal. With ever increasing information flows surrounding the oil commodity, it will be difficult to factor out the externalities at the end costumer part of the value chain. Involved companies and policy makers will probably soon be getting ready for this reality, to take advantage of the opportunity that it could bring, especially for those regions and companies with a competitive edge.

The question is then why this have not happened yet? One reason may be that even though some customers may be patriotic or environmentally concerned, they might also like their petrol to be inexpensive since they spend a lot of money on it.

Another reason could be due to the current industry infrastructure, where it's just difficult to track where the oil comes from, and thus difficult to make sure that the petrol people purchase comes from a certain area due to the large scale logistics of the oil industry. However, this problem could be mitigated by certification schemes similar to that of green electricity or green gas, where the chemical molecule is not the important part but rather that the overall input to the system being impacted.

A more probable reason of why it haven't happened yet, might rather just be path dependence. Where the industry has simply just functioned in a certain way for decades, and have had too many other things on their plate in terms of security of supply, public opinion and renewables to even start thinking in decommoditisation terms. It has also in the past been a quite lucrative market running business as usual, so the economic pressure to change has also been rather limited.

However, due to the rising pressure from information flows, the newborn challenge from renewables and public concern of global warming—it is quite likely that this commodity will go down a differentiation path as well. This development is likely to be underpinned by the huge importance of the oil in terms of geopolitics and environmental impact, as well as the enormous volumes being pumped through the global economy and the subsequent vast revenue streams that could be unlocked by decommoditisation for certain regions and companies.

Since information transparency of oil production are bound to increase, it will thus be pivotal for companies to clean up their act wherever possible and to market positive externalities in order to retain their market shares. Albeit the the days of oil should certainly not yet be counted, the days of commoditised oil might be. It will therefore become ever more important to sure up decommoditisation revenue steams from the commodity as conventional margins simmer away. It could thus be very interesting to see how the oil market evolves in the coming decades, and if information flows will decommoditise the archetype commodity.

7.3 Steel, Metals and Gravel: Will the Historical Commodities Be Able to Defy the Trend?

Another possible decommoditisation in the pipeline, could be the one of steel and metals in general. Steel is the backbone of many advanced industrial economies, and it usually strikes a nationalistic string whenever the industry is threatened. An example of this was in the US so called 'rust belt' which had faced years of decline in steel production due to a number of different factors. Due to the decline, the US government tried to revive the industry in the early 2000s by imposing import tariffs on foreign steel, with the purpose intent that the US would once again produce its own steel for domestic industries. The action arguably made very little sense from a macro economic perspective, where it would be more economical for the US to focus on its high tech competitive advantages for exports and import cheap steel whenever required. Nevertheless steel was considered an important key industry and had to be supported.

A more recent example was that of the 2015–2016 oversupply of steel in global market, where many western nations accused China (possibly rightfully) of dumping steel on the global market below production costs. The background for this was the slowdown of Chinese economic growth, as well as the reorienting from investment led growth to consumer driven growth in the rapidly evolving economy. A situation which reduced demand for steel while production volumes were being maintained at constant levels, thus causing a steel oversupply in the middle kingdom's home market. Furthermore, since the Chinese market was such a large part of the global market, this situation caused an oversupply of steel in the global market as well, and caused prices to plummet for the commodity.

Regardless of who was to blame, the lower international prices for steel and the subsequent closing down of unprofitable steel mills in Britain, caused an uproar from national politicians in the UK. Steel had after all been a vital part of Britain since the industrial revolution and were supporting a lot of jobs. In response to this, local authorities such as Manchester signed a charter for 'Sustainable British Steel' in order to spur domestic produced demand, and the UK government described steel production as a strategic asset which it would examine all options to retain. Some politicians explicitly sited the need for having domestic steel production capacity in order to build war machinery, such as ships, planes and tanks. Britain would therefore, according to proponents, be vulnerable without basic steel production from a strategic military resource standpoint.

This can of course be argued, but does not make that much sense since Britons do not anymore live in the ironclad era, and the production of weapons today is a rather global enterprise. Where much of the British weaponry is already being produced in the US, Sweden, France and Germany. It is also likely that Britain would be able to buy steel on the international markets right up to the supposed war. Furthermore, due to the very short time period future wars against potent adversaries would be waged, the war would probably be over well before it's possible to even reach a final investment decision on a new steel plated ship (which then might take several years to build). War material are in any case a tiny part of the overall steel demand, and if becoming squeezed the country could potentially just build less reenforced steel buildings for a few months. Lastly, a modern war ship would need a lot more than steel to become operational, and a larger worry would probably be access to modern technologies such as IT systems, radar and components for guided missiles which are usually secured through cooperation between countries and bilateral trade.

Nevertheless, steel seems to tempt the imagination and hits a nationalistic nerve in many countries where production is being threatened. Although one can argue the rational regarding the national importance of steel production, in a market economy the customer s always right—and it seems as if many consumers actually prefer and value steel which supports their domestic industry compared to foreign produced bars. This could in itself be considered a partial decommoditisation, where British steel to some British people would be worth more than Chinese steel. Finished products and buildings using steel could thus become certified by origin, which might raise margins for end product assemblers. For example, a nice black Bentley crafted out of British steel might entail a greater value for some consumers than a nice black Bentley crafted out of Taiwanese Steel.

However, the UK as an example has experienced a long terms decline in steel production and lost about half of its volumes in just under two decades. In 1875 the UK pig iron production accounted for almost half of the worlds total output, and almost 40% of global steel production. Today annual global steel production hovers at around 1600 million tonnes, with the UK production contributing less than 1% of

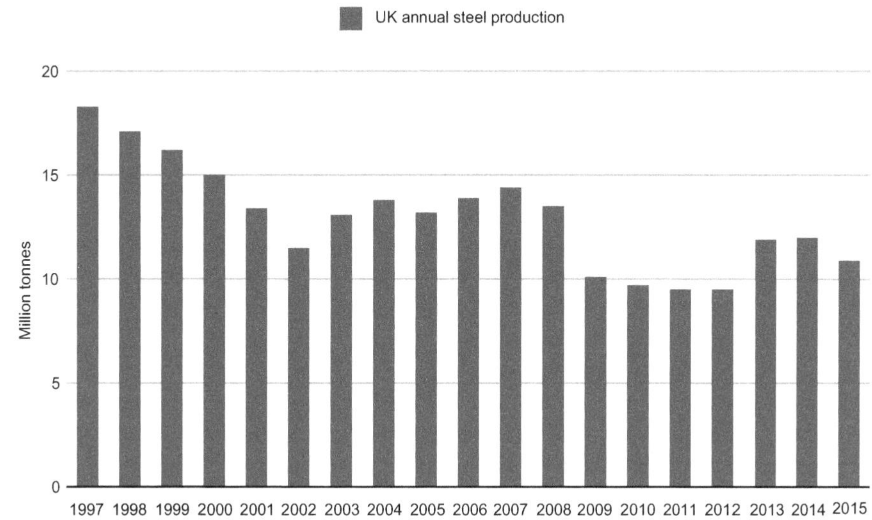

Fig. 7.3 UK Annual steel production volumes. Source: World Steel (2017), Global steel production

global supplies, while China produces around half of global volumes at 800Mt per annum—as seen in Fig. 7.3.

The tough reality is that traditional steel producers such as the UK ones seems to be unable to thrive in the current market and has seen a falling trend in market share for decades. These producers might thus need to shift towards a new business model with higher levels of value added, or else be forced out of business by lower cost alternatives from other countries. Current steel production is also rather carbon intensive due to the burning of coke (high purity coal) in order to reduce the iron oxide which it is made of unless its recycled steel. Approximately 750 kg of coal is used to produce a tonne of steel, in addition to the energy requirement for the whole process which could also involve coal combustion for power purposes. Other decommoditisation aspects part from the nationalistic ones, such as environmental impact of production could thus begin to matter as well. Where CO_2 emissions per tonne of steel, open cast mine versus deep mining, mining in sensitive ecological areas versus mining in gravel deserts, treatment of workers, etc, could also begin to matter for many consumers if being properly communicated.

In 2016 there was already a push from the Swedish government and domestic steel producers; trying to decarbonise national steel production by utilising hydrogen gas produced from renewables, instead of using fossil fuels to reduce the iron oxide. The entailing emissions being water vapour instead of carbon dioxide.

Part from improving the country's carbon footprint, the steel would hopefully also be able to fetch a higher price at international niche markets due to its green profile. In addition to creating a national competitive advantage for the future, there might also be a value in exporting green steel knowhow.

It would not be difficult to imagine similar decommoditisation trends for copper, tin, gravel, and stone as well. The certification schemes might become standard in order to sell the products at a decent margin. Furthermore, other more fringe commodities such as sulphuric acid could also be decommoditised where the consumer through certification schemes are ensured that the inputs to their end product have been produced and recycled in a sustainable way as well. Although commodities such as industrial solvents will probably find it more difficult to decommoditise at an early stage as they do not make up a large part of consumers externality concern just yet. We could expect to see a development of decommoditisation throughout these fringe markets as well if information regarding spreads, and consumers starts to take notice.

Future possible decommoditisation of oil, metals under other commodities which we have examined in Chap. 7 is currently nothing more than mere speculation. It is very difficult to make forecasts, especially about the future—but with ever increasing information density in products and the subsequent higher awareness of the consumers, it will be difficult for any commodity not to be scrutinised on their externalities. So if producers, traders and countries wish to stay competitive in currently undisturbed commodity markets, they should consider differentiating commodities to increase value added and gain a competitive edge vis-a-vis competitors. Otherwise it is quite likely that others will, and market forces can be brutally efficient in the long run.

In Chap. 8 we will explore how countries and policy makers could react to best take advantage of the decommoditisation trend and opportunity.

References

Included with permission from the Natural Resources Defence Council NRDC (2010) Lifecycle emissions of tar sands & GHG emission factors for high carbon intensity crude oils. https://www.nrdc.org/sites/default/files/ene_10110501a.pdf & https://www.nrdc.org/sites/default/files/ene_10070101a.pdf. Accessed 10 September 2017

US Energy Information Administration (2017) Petroleum and other liquids data & analysis. https://www.eia.gov/petroleum/. Accessed 10 September 2017

World Steel (2017) Global steel production. https://www.worldsteel.org/media-centre/press-releases/2017/world-steel-in-figures-2017.html. Accessed 10 September 2017

How Could Countries Reap the Benefits of the Decommoditisation Trend

8

As we have seen in the previous chapters, there seem to be decommoditisation and commodity differentiation already taking place throughout several commodity markets. Since the trends driving decommoditisation is arguably accelerating rather than levelling out, countries involved in the commodity sector would be well advised to develop a strategy for this changing future. This in order to best reap the benefits that decommoditisation brings while at the same time mitigating the risks this entails. In this chapter, we will examine why such as national strategy might be needed, and what consideration such as strategy should contain.

8.1 Is There a Rationale to Push the Trend from a National Perspective, and What Are the Political Implications of Decommoditisation?

The challenge posed to policy makers to facilitate decommoditisation is to create a regulatory framework that benefits most parties through the transition, protect the fundamental base of the economy from excessive shocks, while at the same time spur innovation in the sectors to foster long term competitiveness—piece of cake!

Markets arguably usually works best without too much intervention, but the notion of a well functioning completely free roaming market economy where the state does not interfere at all is largely a delusion. Although the state is probably not the best suited actor to set which price different food items should fetch at the supermarket, or to determine where to set up hot dog vendors in the streets. However, free markets would probably not work very well over the long run unless the state provides law and order, as well as an economic framework with regulation that make sure that the economy progresses forward without too many crises popping up along the way.

F. Andrén-Sandberg, *Commodities Pricing and the Bulk Trap*,
https://doi.org/10.1007/978-3-319-72468-3_8

Knowledge Box: The Dutch Tulip Market Crash, Free Market Failure

In 1637 one of the first recorded speculative financial bubbles collapsed. For years there had been an increased interest in tulips imported by the Dutch far east company, where voyages could yield profits of over 400% on the initial investments made. The tulips where very different from other European flowers, and due to the low supply and high demand, the value of the product increased spectacularly. Speculation on further price increases caused a rush for tulips in the market, which drove up prices of the tulip bulbs even higher. Witnessing the hefty price increases, this in turn caused even more investors to rush towards the market, further increasing prices and fuelling the development. Some people in this period invested their entire life savings, their land and their business, and at the peak of the bubble some rare tulips fetched prices equivalent to ten times the annual income of skilled craftsmen, which in turn was equivalent to 49,000 m^2 of arable land, or 26 tons of Dutch butter!

The prices increased up to a certain point when people suddenly started to doubt the value of the tulips. Price thus started to stagnate and then fall, which caused investors to sell of their tulips, which made the tulip price to fall even further and created a sell of panic for all who had invested far beyond their means in the speculative bubble. People lost their life savings to fund ships being commissioned to fetch the goods, sailors died from the voyages and some traders committed suicide when the crash became apparent—and all because of the speculative price of some random flowers. The Dutch might arguably have been better off investing the economic efforts and labour towards some more productive mean, such as building bridges and more windmills, or build dikes to claim more land from the sea where they could herd cows.

Although some argue that there have been speculative bubbles before in history, this is one of the earliest examples of a speculative bubbles recorded and has been labeled the tulip mania. These kinds of speculation happens today as well, as with property prices in Japan which collapsed in the early 1990s (commercial properties in major cities had experienced 300% increase in prices over 6 years), and seem to be part of mass consumers more destructive group behaviour. Due to the financial turmoil these kinds of speculations causes the 'real' economy and jobs, there are plenty of reasons for government to intervene and regulate in order to smooth out the 'free roaming market' for it to actually function better in the short and long term.

If governments had not intervened at all and let the market roam totally free, societies would probably still have open slavery, there would be a lot less forests resources to fuel the economy without reasonable replantation laws, the ozone layer would still be receding (causing significant increases in cancer prevalence for all species), we would probably had exterminated a bunch of wale species, and in general the global economy would have had a lot more financial turmoil and human suffering due to poor governance of the economic system.

Thus, if we would deregulate all markets and cut all the so called red-tape, we would probably end up with even more Newfoundland fish stock collapses and a cricket mentality economic system without the ability to create lasting prosperity or GDP levels. That being said, regulators have been terrible at picking economic paths for the economy in the past, and is perhaps not the best suited actor for determining the exact route for the economy or for that matter to decide how to best respond to new market conditions. There are countless examples where an industry in stagnation have been protected by governments, just to be swept away by fundamental economic and technological forces several years later at an even higher cost. In these cases regulation has just served to protract the transition and held back the economy from innovating in new economic areas. It is thus important for policy makers to promote a fine tuned regulatory balance, in order to best promote transitions in an orderly fashion without intervening to much in the detailed developments.

Knowledge Box: Swedish Forests Legislation of 1903, Improving the Economy's Long Term Prosperity

Sweden which has historically been covered with forest had in the late nineteenth century become a major exporter of sawn wood and other wood products to the rest of Europe, and the forestry industry was a key pillar for the industrialisation of the country. However, a problem which emerged was that the free market incentivised forestry owners and industry to prioritise short term profits over long term profits, and they were therefore depleting the forest stock which they were dependent upon. The industry were simply harvesting too many trees, while planting too few new trees so sustain the business in the long run. The problem was not solved by the industry itself, since a forest replantation would not yield any benefit to the companies for at least another 20–30 years while replanting costed money right now. Furthermore, just as in the early twenty first century, the most important business time horizon was of course the current year or in the best case coming 5 year period, where they had to keep up with the competition and keep costs low.

In order to mitigate this problem, the Swedish government enacted the Swedish forest legislation which obliged forestry owners to replant trees whenever they cut them down. It also set time spans and limits on when forest could be harvested and general rules to ensure long term re-growth of the forest stock, while at the same time ensuring short term viability of the forest industry and a level playing field for competitors.

The legislation proved to be a success, with the Swedish forest stock replenished and saw a 109% increase in tree volume over the following 90 years—ensuring that the forestry industry could have a long term future

(continued)

in the country. Today the Swedish forestry industry products and derivatives generate a significant part exports revenues and is a vital part of the industrial base. The episode exemplify how well balanced regulation can be beneficial to long term economic prospects for a country, where intervention can smooth out the 'free roaming market' for it to actually function better in the short and long term.

In terms of the decommoditisation trend, the market seem to be doing lot of the work by itself, but there are several instances of market failure where regulators are well suited to intervene to foster an even quicker progress in certain sectors.

In terms of origin certification of goods, there have for instance been problems with fraud and smuggling which the private sector are ill suited to manage. In comparison; trafficking of narcotic drugs is a relative risky business with usual harsh punishment if being caught, and there are typically government backed agencies and specialised task forces handling the issue. To smuggle counterfeit food on the other hand, where regular farm products are being faulty labeled as organic and sold at much higher margins, can also be a very lucrative enterprise but is currently next to impossible for customs to detect. It is also difficult for retailers of the goods to know for certain if the Romanian tomato is in fact organically produced, which makes them more or less reliant to just trust the branding or certificate on the packaging.

This kinds of situations must be addressed in order for decommoditisation to take place. If unlawful practices such as these becomes wide spread, consumers may start to doubt the validity of brands or certifications and not be willing to pay an additional cent for it. In the case of the food example, this development could for instance harm honest organic producers with higher production costs and could force them out of business. This would be a very dangerous development for those attempting to differentiate their commodities, and could potentially derail or at least postpone decommoditisation across an entire market, or even across several commodity markets if becoming too widespread.

It is therefore pivotal that governments and regulators intervene at an early stage, setting up tools and enforcement practices which combats the dubious actors in the market, and disincentivise their business models. Exactly how to do this will depend on the specific commodity circumstances, while also be up to each regulator depending on national economic model setup and the country's general institutions. It's is furthermore likely that the regulation will have to evolve over time as technology and the market changes. Regardless of the specifics, collaboration across borders will be key for an efficient trade to take place, and it may even be necessary for regulators to run their own certification systems themselves if private actors are unable to credibly fill the void.

Legitimate states may enjoy a high level of credibility, and clear certification schemes for some goods run by governments could therefore be an option, but one must be mindful not to over regulate and to confuse certification schemes with lawmaking. Laws governing the basic rights of for instance workers and farming practices are already run by governments and should constitute a lowest possible level of responsibility. Certification systems on the other hand does not outlaw non-certified goods, it only ensures that the products or services reaches certain standards stipulated by the certification contract. The regulation should neither be too cumbersome that it risks strangling innovation in the sector, and must be flexible enough to encompass fairly quick changes along the journey. A possible solution could be government and non-profit NGO partnerships towards certifications, where certification agencies establish and adjust the rules in a dynamic manner, while governments make sure fraud is kept at bay through strict enforcement. Government certification is in fact today present in several energy sectors in terms of green gas certificates, green power and carbon emission rights, but also within pharmaceuticals which is a very carefully regulated market. So, there should already be well established system set-ups to draw experience from if wanting to expand certification towards new product categories and markets.

With even increasing information density in products, technology might also be able to relive the validation problem, with clear linkages from source to consumer with hard to fraud certificates and unique ID encryption keys per product unit. This would enable consumers to easily check whether their specific goods has been produced in the claimed manner, and if the specific product certificate has already been used or not. This makes it possible to ensure that a cucumber bought in a German supermarket has in fact been produced in Spain with organic fertiliser and transported to Austria by electric vehicles, driven lorry by a driver having decent working conditions—all insured by the tracking tag.

Crypto currencies such as bitcoins already utilise blockchain technology to tag all transactions with a distributed ledger—and similar technologies could be well adapted to track the production and consumption of goods. Although it may seem excessively costly to attach specific IDs to all goods produced, but considering the pace technology evolves and are becoming cheaper, this could in the future actually be a rather viable and cheap option for some current commodities. For packaging, an added bonus would be the possibility to keep track of recycling volumes, making sure that most of the produced packaging is reaching its intended destination to close the recycling circle.

Regardless of the specific technology solutions, it will be very important that governments realises the potential in decommoditisation and the economic benefits that well balanced regulation may bring.

8.2 Understanding the Risks of the Decommoditisation Process

Decommoditisation might be far from a risk free enterprise. As we have seen in the energy sector, decommoditisation have turned market leaders close to bankruptcy—and could in the same manner bring wealthy countries to the brink of default in the if they do not adapt to market conditions. Oil producing nations like Kuwait without significant oil revenues might for instance find itself in quite a tricky economical situation, if not having changed their commodity strategy.

The energy industry may be the most vivid example of decommoditisation impact due to it's relative large share it inhabits in modern economies, and it is unlikely that other single decommoditisation processes would yield similar upheavals. But as many sectors might be going through decommoditisation at the same time, it may add up significantly. Furthermore, since many smaller nations are reliant on just one or a few commodity exports for their income—insufficient regard for decommoditisation may entail substantial risk the countries fundamental economic wellbeing.

Since policies enabling decommoditisation will be key to unlocking the economic potential of new industry niches, countries who strikes the right balance might gain a considerable upper hand vis-a-vis those who misinterpret the developments. This also implies that the countries who gets the balance wrong may be outcompeted by peers, or may experience economic suffering due to poorly managed investments.

The decommoditisation main risks for nations can thus be categorised into four main sections:

- Slow movers and no movers, including those who believe that the market will solve everything. These are unlikely to reap the early benefits of decommoditisation
- Those who move too fast, who might suffer substantial economic pain without reaping proportional benefits
- Those who move at a lagom (modest) speed, but fail to implement efficient structures to reap the benefits
- Micromanagers who tries to steer the exact development of the market by excessive regulation. They will likely fail since lacking flexibility to adapt to evolving market conditions

These risks should furthermore not be seen as something that mainly affect commodity producing countries, although the risk for these might be especially large if their economies are not sufficiently diversified and mainly dependent on commodity exports. However, the risks also applies for all countries which are present in any commodity value chain, namely all countries!

Product and service providers which require commodities as raw input for their industries, as well as retailers of finished goods which uses primary commodities are also susceptible to these risks. The retail sector is a major part of most advanced economies, and retail can also be a great source of income as companies such as IKEA, Walmart and H&M has shown. Companies such as these might easily be outcompeted by their peers if failing to react to new market conditions and commodity requirements from their end consumers. If companies such as these are not exposed to sufficient regulation in their core markets, they might also be unable to compete in other markets with tougher regulation. However, market actors exposed to progressive regulation will adapt and eventually start to thrive in their home markets. These companies might then have an edge in exporting the new products and services into markets which do not have the progressive regulation, but where consumers have picked up the trend anyway. The unregulated incumbents may in these cases have a significant disadvantage stemming from the late start, which might prove to be a major differentiator in the long run and cause them to loose market shares to the newcomers from stricter regulative environments.

A solution for businesses stuck in unprogressive markets could of course be to pick up on the trend anyway and thus remain competitive, but will likely be hesitant due to the risk of higher costs in the short term compared to peers without certainty of future market. It will therefore be vital for countries to spur the trend to gain a first mover competitive advantage, and to create progressive regulation to enable this initial growth momentum

8.3 Design of Policies to Promote Beneficial Decommoditisation

It will be important for policymakers to create flexible regulatory regimes where innovation can occur, while at the same time ensure sufficient rigidity for consumers and market actors to trust the long term viability and credibility of the system.

The design of certification systems will likely be very important, but will differ depending on the characteristics of the commodity. If there is an actual benefit of the differentiated physical commodity (even if far fetched), the specific commodity unit and certificate need to be directly attached. An example of this would be an organically produced tomato, which beside the 'good' production methods might also be deemed 'healthier' to some due to the absence of pesticides, increased nutritional value, etc. It is therefore important that the specific organically produced tomato end up at the customers dinner plate, and not just a general tomato part of a batch where a certain quota of organic tomatoes have been introduced.

In contrast, if what is important is the overall global commodity volumes composition, then there is no reason that the specific physical commodity the customer consumes is the actual one produced in a certain way. In this case it would perhaps be better to trade certificates and the products separately, where the certificate in not necessary attached to the specific physical commodity itself. An example of this would for instance be green gas, where renewable methane is injected into a natural gas grid. Since it's impossible to keep track of the specific gas molecules, the given amount injected just changes the overall composition of the gas and then proportionally produces green gas certificates. These certificates can then be bought by consumers on the other side of the gas grid, who despite not getting the actual 'green' methane molecule that was injected, can still rightly claim to be consuming green gas since they contributed to an overall increase of green gas in the system by purchasing the green certificates.

The reason this would be a preferred option for green gas, or similar arbitrary commodities, is that the alternative would be to build up separate distribution networks or other logistical systems. This would make the end products both more expensive and polluting without any entailing additional benefits for the system. The methane molecule consumed will from a technical standpoint be exactly the same regardless if it's renewable or not, and the only thing that matters for the end consumer is thus the externalities caused by its production. The supply and demand of the certificates will then determine the input mix of methane, thus forcing out fossil alternatives while increasing the renewable part (or benefitting domestic production, or other trait depending on what is being differentiated). This type of system would be well suited to support decommoditisation of metals, energy products, oil and other goods where the differentiation does not matter from a technical utilisation perspective. However, it would be a badly suited model for commodity inputs such as clothing fabric, food items, or plastics where the end product composition actually matters for the consumer.

Certification of origin is another sort of certification, where the certificates themselves cannot be treated as commodities at all. The reason for this is the further differentiation, where it will not only matter in which category manner the item has been produced by, as with general renewables, but also the precise methods such as origins, technology, specific emissions, etc. Another example of this would be consumers buying food from a certain farm using very specific production methods such as insect pest control to get their carrots to market. Another example could be power being purchased from a specific wind farm at a certain place, or solar power from a specific neighbour. Peer-to-peer blockchain technologies is already being developed for the latter by some companies, where consumers can trade power amongst themselves.

However, in order for this to work, authorities may need to have processes in place to police origin claims and similar; in order to limit unserious actors popping up and disturbing the processes. That is unless the markets can solve it by itself, by utilising blockchains or other self policing technology.

Another complementary way to introduce decommoditisation is through rising quota systems. In these cases a certain amount of a goods trait, such as CO_2 free steel, is introduced into the market by law—stating that a certain share of the end product tonnage needs to be sourced from a certain criteria in a certain year. For example all consumers of steel are required to purchase at least 4% steel produced from CO_2 free processes in 2023, rising to 6% in 2026, rising to 20% in 2030 and so forth.

This will ensure a solid market for the niche producers and reduce market risk for purchasers, as supply and demand should ensure that there will be products available in the market. In order to deal with initial scarcity, a window out can be created by stating that unless quota is fulfilled, a certain amount of money needs to be payed to the government as penalty. This also effectively caps the maximum production cost for the differentiated commodity, as the alternative cost is just to pay the penalty if the market cannot produce under that cost. The regulator can then also choose to earmark the additional penalty income to further spur the industry, by funding demo projects, research or similar to create a solid loop into the system.

An alternative to create even more solid transformation loops is through bonus-malus systems, where the conventional suppliers are made to pay incentives for the new entrants. For example, fossil fuel cars are obliged to pay an additional 10% tax on purchase which will be used to subsidise the purchases of electric vehicles. The benefits of these systems is that they are largely self regulating in terms of volume, where a 5% electric car market share will get huge incentives from the other 95%—which should make the market grow until a market balance is achieved. As costs for the new goods comes down in price due to the larger volumes, less subsidies should be needed and the market share can thus further grow until it takes over an even larger part of the market. When the fossil fuel vehicles in the end only comprises 10–20% of the overall market, their contribution to the transformation will have become rather limited—but has in that case already achieved the overall goals. An alternative is also to raise the contribution share, from 10% to 15% a few years later and so forth, in order to almost completely phase out the unwanted commodity traits.

Success breeds ambition, but ambition might also breed success

This kind of quota systems will be suitable to commodities, which will not differentiate too much and are quite binary in their externalities, such as renewable and non-renewable, or hydrogen blended with natural gas. On the other hand it is badly suited to spur innovation and development in a sector with a lot of diversity and spectra within, such as with food production. In the case of organic farming, quotas and goals can be set us as many government has done. But since organic farming practices will vary quite widely in their externality impact, is might just cause another price race to the bottom where the most cost efficient organic solution wins the day. This may live up to the least minimum denominator, but may not yield the substantial benefits initially promoted and stifle high margin niche growth. As an example; organic farming can be a local enterprise or shipped from the other side of the globe, it can involve factory farmed animals who eats organic fodder or free range varieties eating grass on a savanna. With substantial variety and innovation possible which will not be promoted by fixed quotas, it might even be a factor strangling burgeoning niche sub-markets. That is unless the quotas just sets the base floor, and the purchasers are then free to pursue full spectra of differentiation.

Governments can also choose to set up rating or information system, similar to those present for energy declaration in the EU; where a label is attached to all power consuming appliances as well as buildings, which ranks their energy efficiency in terms of a certain benchmark. With the design intent to help consumers choose according to the preferences by simply mandating information sharing. Similar information requirements have already been attached to food items, where producers in many countries now needs to state the calorific and nutritional content of the products. Information akin to this could also be made mandatory for carbon emissions, where all food and other goods needs to state their greenhouse gas impact. This could have a significant effect on the decarbonisation and decommoditisation of primary products, as customers becomes better informed regarding the externalities of their purchases at the time of purchase.

Another mandatory option would be that all products needs to state exactly where it was produced. Not only generally such as "Made in Taiwan" but also at which exact factory, farm, workshop, or similar—including primary inputs. All this information might be difficult to attach to most product packaging, but could be condensed on the label and then being fully available online through a QR code (2D bar code) or similar.

The drawback with mandatory information stipulated by governments is that it adds administrative costs, including the requirement of tests and oversight of the industry. Regulators should therefore be mindful not to over-regulate and excessively burden the suppliers with draconian measures, which in itself might just increase cost to the system without necessarily bringing the intended benefits.

If wanting to introduce a system through government tenders, a good solution could be official tenders where a specific quantity are requested for a certain price—which will introduce fierce competition between suppliers to bring down costs. This system have worked well for the energy industry, where volumes are being bought up by the state, but might be less useful in procuring sustainable wheat or similar more diversified goods. The design of well functioning auctions system, will be very much dependent on the specific circumstances, as well as the intended outcome—and is probably a book in its own right.

If just wanting to introduce a new decommoditisation technology which might yield substantial disruption in the long run but are currently unable to compete on market conditions alone, then a direct subsidy might also be needed. The market failure at place which could be alleviated in these cases, is that neither market actor has sufficient market share and cannot afford the long term cost of bringing down the price of the production method themselves. Also if a brave company would go it alone, then some other actor might just use their technology once it becomes cheap enough without having to pay for the initial development costs, unless it can be sufficiently patented. The solution to this could be a feed-in-tariff system, which ensures that all early movers will have profitability on their investments and thus dares to commit to the sector. This can be supplemented, or predated by demo project funding, seed finance, grants, or research funding to create an initial supply to the market and test new methods and technologies.

Which of these initiator options to utilise will be very much dependent on the specific circumstances, and especially the maturity and barriers present in the market. If the industry seem to be solving the market imperfections by itself, there will be less need for intervention and vice-versa. However, even good industry solutions could be further spurred by government commitments, which may yield substantial competitive advantages in the long run. In the case for MSC branding for fish, government could for instance easily spur the trend by deciding that all public purchases needs to fulfil these demands. The issue might in this case be that the government is favouring a certain certification while neglecting others, which could hamper competition and efficient innovation. The government should rather set up own certain criteria which needs to be fulfilled for their purchases to spur the industry, and let the market itself choose the most efficient form of certification. Its also very important that the authorities do not confuse law making with certification. Lawmaking are supposed to set the lowest possible standard for goods, whereas certification should be used to assign a premium to a product, which is well above the stipulated minimum requirement set by law.

We have now examined how countries could respond to the decommoditisation trend, and what considerations that needs to be taken when and if deciding to promote it. But what about the industry itself, how should they respond and what could be the natural strategy for different parts of the value chain going forward? This is what we will examine in the following Chap. 9.

How Businesses Can Become Winners of the Decommoditisation Trend

9

Countries seem to be able to benefit from the commoditisation trend if they position themselves with the right regulatory regime in place. But the main 'actuators' of the decommoditisation will nevertheless be companies which service the new demand stemming from the consumers. In this chapter we will examine what aspects that needs to be taken into consideration by those different market actors to become successful in the new differentiated landscape.

9.1 Winners and Looser of an Evolving Commodity Landscape, and the Strategies Which Might Set Them Apart

Is difficult to forecast, but as awareness of production externalities grows it's fair to assume that the decommoditisation trend will continue, and in order for companies to take advantage of and thrive in this paradigm shift; solid strategies needs to be ironed out.

It's clear that this trend will have both winners and losers among companies, and just as with countries; the losers may likely be the ones that both move either too fast or too slow. Moving too fast could mean that the company will not have any demand for the new great differentiated products, while moving too slow might mean the same thing for the outdated commoditised offerings. In the transitory period the golden path for most companies should thus be transition at a 'lagom' (good enough) speed—striking a fine balance between transition and reaping solid revenue from conventional business activities. It might in theory be possible to niche a company in the transitory period towards the old bulk market, by just trying to harvest as much revenue as possible in the short run—but the company will then run the risk of missing out on greater margins by increasing the value of the goods in the longer run.

A company's worst nightmare is for their goods be to rapidly become obsolete, and the type of companies which this regularly happens to are quite often the

F. Andrén-Sandberg, *Commodities Pricing and the Bulk Trap*,
https://doi.org/10.1007/978-3-319-72468-3_9

previously most successful ones. The business graveyards are littered with companies who did very well at what they where specialised in. Maybe just a bit to well, thus not being forced to innovate or change, and once they did; it was either too late or they did it too marginally for it to sustain their business.

Knowledge Box: Nokia Cell Phones, from Leader to Bleeder in a Few Years Time

Nokia used to be world leader in mobile phones, but miscalculated the development in the industry towards smart phones and tablets. Since it was used to be the world leader in mobile phones it developed its own platforms and tried its best with its Symbian operating system, but they were outcompeted by IOS and Android devices from Apple and Google.

Since not wanting to succumb to the competition, Nokia eventually entered into a partnership with Microsoft in 2011 to use the Windows phone platform for its mobiles. However, the strategy did not turn out very well and the whole Nokia mobile phone division was bought up by Microsoft in 2014 as an attempt for Microsoft to integrate the value chain for it's own operating system.

The deal was in hindsight seen as a terrible investment for Microsoft who bought the company for USD 7.9 bn, and just about a year later had to do an impairment of USD 7.6 bn on the acquisition. With thousands of employees loosing their job and the company going from being the worlds largest mobile phone producer with a peak global market share in 2007 of over 40%, to more or less gone in under a decade, it just shows that even huge successes can be very fleeting if not properly adjusting to new market conditions. The decline even had a significant impact on the Finnish economy, where Nokia used to be a large source of exports revenues and a local buyer of consulting services and other things that fuelled the economy.

As a note: The mother company Nokia still exists and has among other things focused on telecommunication infrastructure, and has arguably been rather successful in this area. Nokia also launched an innovative cell phone in 2017, so might also be that this story needs to be rewritten and put into another context in a few years time.

There is always the risk of trying to breed a faster horse instead of inventing the car. Likewise, a mouse trap company who tries to innovate may just end up spending a lot of resources increasing the efficiency parameters of the mouse trap, just to be forced out of business by a company who invents efficient mouse poison.

> Success is a poor teacher, because it seduces us to believe that we cannot fail

Companies who wants to stay ahead of the pack must therefore innovate, and do so from a customer demand perspective. In a decommoditisation context, it may mean inventing a new market niche consumers did not yet know that they wanted, or just follow the path being set out by consumers and reinforced by NGOs. As information flows and transparency increases, companies will become less able to angle information to suit its own purposes, and they might severely loose credibility if being perceived in doing so. Hence, the best course of action may be to 'go with the flow' while relying on external parters to drive the trend forward, and just supply products and services that fits the current market demand if unable to steer the market in any meaningful manner.

To succeed in this new world, strong brands will be a key success factor, and building them will require thorough work. An alternative to building strong brands that supports differentiated products is to instead rely on certifications bodies and their brand value. However, a good strategy is probably a combination of the two, which also implies that companies may have a rationale to collaborate with the certification bodies to strengthen the certification labels.

It will therefore be crucial not to see companies as lone islands surrounded by walls, but rather to work closely with NGOs, certification agencies and other external stakeholders who might have stake in the value chain. This kind of collaboration enables the creation of an information 'mesh' to encapsulate the consumers and their media channels with common messages, rather than a single line of communication stemming from the company.

Since information about the commodity will be the diffracting factor turning it into a product, communication will be a pivotal factor in any company's communication strategy. Information has to be made easily available, transparent and accountable towards all stakeholders. Hence, data should be published in an easy to understand format and put into the right context to conclude what the impact for the products actually are, and technical data not just dumped into spreadsheets towards stakeholders.

Just as with diseases, the most relevant information to determine outcome is not always the current state of the patient, but rather the trajectory of that state. The same might hold true for decommoditisation communication, where companies won't have wait to convert all production into the preferred manner before stating their intentions and trajectory. If there is a strategy in place with clear goals and a credible roadmap towards getting there, this should be considered to be communicated in order to position the company as being on the right track, which the customers might be willing to support. However, long term storyline should be carefully considered as consumers might quickly loose patience if the company does not live up to their promises on such a roadmap.

Information and communication will be of crucial to differentiate commodities and receive higher revenues, but there of course also needs to be something to communicate. Exactly what this something is, will differ widely from industry to industry and from what customer base that is being targeted. The most common differentiator in decommoditisation so far has been regarding sustainability, environmental and social practices. These aspects will most likely also be key differentiating factors in the future, especially for commodity production that has large impact on the environment or is labour intensive. However, there could also be other differentiating factors such as the geographic or nationality of production (patriotism), the increasing use of humans instead of machinery and robots (job creation), using labour from disadvantaged groups such as ex convicts or people with disabilities (social responsibility), promotion of positive community building externalities, or something completely else. Just as for other factors of products and services, the imagination and customer niches will probably the limiting agent in this regard.

9.2 Commodity Producers: What Could be the Decommoditisation Approach

In order for primary producers of commodities (such as oil producers, iron melters or miners) to reap the benefits of this transformation, the first step would be to determine what negative and positive externalities their production entails. The next step is to work out just how the negative externalities could be minimised while the positive externalities being maximised, as well as determining if there could be other positive externalities added into the production process. The next and perhaps trickier step is to find out if there is market demand for the differentiated product. This needs to be done in close collaboration with partners downstream the value chain—entering into discussion with end customer retailers, traders, wholesale companies, NGOs and even government entities to find out if there is a market out there, or whether a market could be created.

The latter part may be quite tricky, since there in many cases probably won't be a market for the product (former commodity) as it has yet to exist. When for instance customer demand for computer tablets were surveyed before they where introduced, indications where that there were not a very large market for it. However, after being introduced in the market they became a huge success by creating the demand for itself. The same might hold true for differentiated commodities, where consumers simply not yet know that they wish to buy automobiles crafted from steel that has been sourced domestically or using renewable plastic parts in their toys. Hence, it might be that markets in many cases has to be created, and the first who does so might probably be able reap higher margins than late comers.

After the market assessment have been done, investment analysis and business cases needs to be evaluated. These will most likely be riddled with high levels of uncertainty and probably needs to contain some leap of faith regarding the creation of a future market. That is unless contracts can been signed down the value chain where counter parties bare some of the risk, or governments can be persuaded to create partial or full market certainty. However, there's no material difference from this to any other business decision—where managing risks is what great companies and business leaders are supposed to do well. The tricky part in this case will rather be to transform the business at the right pace—not be too fast so the market demand cannot pick up, nor too slow compared to the competitors. The latter consequence being that demand from the original commodity will decrease while competitors gets a head start in terms of new product offerings and market share. Judging from history, these transformations can come fairly rapidly and the pace should not underestimated—where markets can be overturned by escalating economics within just a few years time.

The final step would be to establish a roadmap on how to reach the stated goals, launch information campaigns together with partners across the value chain and to execute the strategy while constantly evaluating the market to adjust in regards to new conditions.

Quick guide to decommoditisation strategy for businesses:

1. Mapping of externalities
2. Evaluate alternatives to 'optimise' externalities (i.e. reduce negative externalities, increase positive externalities and introduce new positive externalities)
3. Assess market potential in close collaboration with partners in the value chain
4. Establish and evaluate business cases
5. Implement roadmap together with partners while constantly evaluate market developments

9.3 Primary Users of Commodities: How to Develop Decommoditisation Strategy from a Manufacturer's Standpoint

For companies using commodities in its value added production such as Samsung, Volvo or BMW, as well as commodity traders and retailers—the approach towards decommoditisation should be somewhat similar to that of producers. Albeit the strategy might be viewed as slightly more tricky since not having control on the production levers, or complete visibility of the externalities involved. The first thing that thus needs to be established is what commodities are being used across the product portfolio, and what externalities the production of these commodities entails. The next step would be to grade the externalities on impacts and ease of implementation (e.g. 2×2 matrix), and to start up discussions with suppliers of potential options, including estimations on costs and time horizon for each alternative. Market assessments then needs to be made, and business cases evaluated to target the lowest hanging fruits first. This means assessing which of the used commodity externalities that has the largest impact according to the customers, and which of these can be mitigated or enhanced at the lowest possible cost—most bang for the buck!

The last step would be to establish a roadmap towards the targets and implement the strategy in conjunction with partners across value chain, including the communication strategy. If it turns out that it's difficult to attain the commodities and products that suits the demand of in the market, these may have to be created in dialogue with the commodity producers. A way forward might be to contract commodity volumes at a specific price to create certainty for the commodity producers, and thus pulling new production options into the markets. If the own volumes turns out to be too limited for covering an investment at the source, it may be worthwhile to even reach out to other users of the same commodity who are not in direct competition with the business (e.g. BMW reaching out to a construction company regarding sustainable steel), and gauge if they too could be persuaded to guarantee volumes. An alternative could even be to co-invest in the production of the the new inputs, thus de-facto entering into another part of the value chain to share some of the risk. Just as with volume contracting, co-investments could also be considered to be done in collaboration with other partners in joint ventures—which might also entail a greater value from a communication perspective. If the implementation turns out to be a success, it would also be relatively easy to ramp up volumes with existing partnerships.

Just as from the producer perspective, the decommoditisation may have to be done in collaborative networks outside the direct value chain, which might include NGOs and government bodies. There could be instances where government supports the trend with subsidy schemes or certification, but in cases where there are insufficient government support; the industry may also have to consider proposing legislation that can incentivise good practices while discouraging less fortunate ones.

9.4 Retailers, NGOs and Political Organisations: Driving Seat or Back Seat in the Decommoditisation Trend?

For retailers of finished products such as IKEA or Home depot—the decommoditisation approach should be very similar to companies using commodities in its production, just one step down in the value chain. They will need to determine what commodities and externalities their products or services entail, cooperate with relevant suppliers and NGOs, or possibly even government bodies. Grading the commodities on externality importance and begin to gauge suppliers possibilities to provide alternatives including estimations of the potential cost increase. If there are limited appetite from suppliers to produce the goods, then there might be a case for co-development in order to share some of the risks. The last step will be to assess the market for the product, restock inventories when appropriate, and launch communication strategy towards the end consumers and other stakeholders.

These companies who actually faces the end customer have a crucial role in the decommoditisation process, as they will have the largest surface area for suspending information towards the customer. This entails substantial possibilities to reap high economic margins depending on the value chain power setup. As these companies are usually also responsible for the primary marketing of products, they may also become the kingmakers by creating demand for certain types of decommoditisation.

Organisations outside the main value chain will also have a vital role in the trend, and NGOs will likely be pushing companies involved in production or distribution to take action. The goal for these entities will be to spur decommoditisation in all parts of the value chain and to raise awareness at the consumer level in accordance with their agenda. Together with the active consumers, NGOs will be in a leading position to change both companies behaviour and customer preferences.

The strategy for NGOs should thus be to map out commodity usage and their externalities for a wide stock of goods, and study which possible options there could be to mitigate these externalities. Since any transformation has to be backed by consumer opinion supporting the change, its important to communicate and reaching out to the end consumers regarding the adverse effects in order to raise awareness within companies.

After mapping and creating awareness, the next step would be to showcase alternatives to the current practices (assuming deemed not satisfactory) to create market demand for new product offerings and pathways for companies to follow. Another possible step will be to create certifications for a commodity, or group of commodities in order to supply a choice for consumers who still wants to use the goods. As previously mentioned, the certifications does not have to regard sustainability, but could as well be national certification of origin if that what the consumers are asking for.

The final important step for NGOs will be to continuously engage companies to inform them about the possibility to change, raise awareness at the consumer level and thus create the subsequent business rationale for the transformation. An important angle might be to make companies understand that it is not money incineration to enter the decommoditisation transformation, but rather an opportunity to increase margins by adding value to their commodities or products, and to escape the bulk trap where they might currently find themselves in.

In parallel to company and consumer engagement, there could also be similar NGO campaigns towards governments, politicians and regulatory bodies to create awareness and inform about their opportunity to support the trend.

9.5 End Consumers, Voters and 'Social Media Dragoons': The Actual Gatekeepers of the Trend

End consumers will be the main driving force for the whole decommoditisation trend. Just as consumers pushed the commoditisation trend by demanding low cost mass produced products to supply their basic needs in the last century, consumers will certainly be pushing the decommoditisation trend in this century as well. For general customers who wants to push this trend forward, the main thing they need to do is just to vote with their wallets—purchasing the products that meets their specific needs and supports their beliefs on how the production and distribution of the goods should be conducted.

The political system in countries are quite important in determining societal outcomes, especially in functioning democracies to support change that the public wants. However, consumer power may be just as, or even more, important in the long run, and also has the inherent benefit that its partly detached from most political systems. If choosing to utilise the taxi-service Über, people automatically tell the market that the future of transportation should use this business model, and that this is the world they would like to live and work in. The same holds true choosing Airbnb compared to general Hotels or electric vehicles versus diesel cars, and the market will almost always effectively respond to the current as well as the forecasted consumer demand. The multiples choices made in the billions by consumers everyday is arguably the strongest force deciding what and how things should be produced, what will be put into R&D and also what companies should focus their efforts towards.

Very engaged consumers can of course also team up to form consumer organisations, push forward legislation through the political system or run agendas through NGOs or companies. However, the sum of everyday small choices has a remarkable strong impact on the future, and might be even more important than these more far reaching engagements by some consumers.

Vote with the wallet, persuade others to do the same, and be early adopters—Engaged consumer ethos

Another important part of the consumer engagement will also be to communicate what products and services are preferred to their peers. This in order to spread information regarding production practises and the alternatives to other potential consumers of the product or service. Peer-to-peer advice from friends and family will likely be more highly valued than corporate information, and with the onset of social media such advice can spread extremely rapidly. For instance, if one person posts their purchasing opinion towards their 250 Facebook friends, and 20% of these repost the opinion to the same number of unique friends—this occurrence will only have to repeat itself 5 times in order for the post to have been reposted over three hundreds millions times and seen by many more. Even with less unique Facebook friends and lower percentage of repost (which is likely), the virality can still reach impressive numbers and are also valid for other social media channels such as linked-in or twitter.

Customer communication to companies, governments and NGOs may also be fruitful to inform what products is being demanded in the future. So in short, the role for everyday consumers wanting to drive decommoditisation and would simply be to:

- Demand products they want from the companies they purchase it from
- Inform family, colleagues and friends about the choices made
- Support organisations who's brunting their cause

These few simple actions may ultimately be the ones that completely transforms the commodity logic of the future, and with social media as a tool—the true future power will very likely lie in the actions of a few influential people starting viral messages which affect the global market in ways we could not yet comprehend.

It's seemingly not very hard for business to notice that decommoditisation is ongoing and also accelerating though several commodity markets. The difficult part for business though is that the future is by definition unknown to all of us. Market actors thus need to strike a fine balance to not missing out on a trend that will revolutionise the market, while not rush towards a trend that will never take hold—as we've seen in this chapter. The question though is if decommoditisation really is an unstoppable escalating trend, or if there might be something on the horizon that could disrupt the trend. In the following Chap. 10 we will examine what possible derailers that could halt or even reverse the decommoditisation trend, and the likeliness that these will occur.

Possible Decommoditisation Derailers 10

In the previous chapters we have followed a red line of arguments which leads to the conclusion that everything will eventually become decommoditised. It is somewhat likely that many commodities will be decommoditised, but there are always market forces beyond our knowledge horizon and comprehension that might counteract this forecast. In this chapter we will examine three alternative scenarios that might unfold which could disrupt decommoditisation, and what signs market actors should be aware of.

> Everything comes in threes, you just have to wait for the third one to occur

10.1 Information Overload: Who Has the Time to Care About the Origins of Goods?

A recent observed trend is that the general population as well as politicians and regulators are being overwhelmed by the sheer abundance of information available, and the opportunities the information represent. In the past, most consumers did not have to worry about which energy company or power option to choose, they just used the one that was available locally and trusted that they were running the operation in the best possible manner. Same was true for many other goods and people generally did not have to worry about how their products were being produced, since the government were supposed to make sure that the goods were safe to consume and being produced in a law abiding manner. Most of the products were furthermore produced either in the consumers own country, or a neighbouring one with similar labour and environmental standards.

In contrast, nowadays consumers are being overwhelmed by choices. In the supermarket they can choose between ten different brands of toothpaste; all with its own subcategories whether you want to have clean teeth, white teeth, extra white teeth, anti coffee stain formula, or with eucalyptus oil to smooth your gums.

F. Andrén-Sandberg, *Commodities Pricing and the Bulk Trap*,
https://doi.org/10.1007/978-3-319-72468-3_10

The same holds true for most products, and it becomes increasingly difficult for companies to compete for consumers attention span. Therefore, even as decommoditisation will be ongoing within the food industry, consumers might experience that although there is an abundance of new information of the products available, it's becoming increasingly difficult to know what is the right choice. Should environmentally conscious Elsa from Norway purchase the organic free range hens from Argentina, or the non-GMO local produced cage hens from the farmer just a few clicks down the fjord? If she looks for some helping advice online, she may also find it quite difficult to assess what is true and what is not, and which aspects of that potential truth is the most important ones. With diet advice, it's possible to find equally compelling storylines on the internet that people should eat food only containing fat and proteins, as one can find storylines stating that people should do everything they can to avoid consuming the same specific carbon constellations. Both storylines complete with background science material endorsed by professors, doctors, scholars, experts within the industry and a few famous celebrities who can assure you that its all true. Which storyline to choose from?

> The classic historical power divide has between those with access to information and those lacking access to information. The power divide of the future will be between those who can browse through the abundance of information to find whats relevant, and those who cannot

Furthermore, the chicken is only one of the choices busy Elsa need to make this day; shampoo and conditioner is down the next isle, rice and pasta after that. Not to mention how to best help the climate when transporting herself to the store, which heating to provide for her hut, which insurance to buy for her sister Anna, what bank to store her savings in and which kind of account—and now the market also expects her to start caring where the oil and wood actually comes from? There may simply not be enough hours of the year available to sift through all information needed to make informed decisions on each and every purchase while also maintaining a functioning life in parallel. Frankly, there may also be quite a lot of other interesting things in life to care about, than the evolving differentiation of commodities

> "I spent a lot of money on booze, birds and fast cars. The rest, I just squandered"—George Best, Northern Irish professional football player (who perhaps did not cared so much about what steel his fast cars were produced from)

As the consumer attention span simply is limited, this might prove to be a substantial problem for decommoditisation. It might be the case that the solution for the overwhelmed consumer will be to simply distrust the whole thing, and just don't care at all. After all, it may all be just greenwashing!

The solution for businesses to overcome the information overload problem will probably be to reduce the number of choices the consumer have to make by categorising and bundling purchases. Instead of figuring out what products are the most suitable to buy individually, the consumers may wish to just make a few simple choices regarding brand and certifications instead. For example; if the fish consumers purchase has the MSC logo on them, they don't need to become an expert on the fishing industry—but simply just learn what MSC stands for. If they trust the MSC certification and its inherent qualities, they then chose products with that label. The same might hold true for brands and outlets. If Apple is trusted that all their products is produced in a good manner, then consumers can just buy products with the bitten apple logo and even showcase it as a badge of responsibility. If all of IKEA products are deemed as produced under ethical conditions, then consumers can just show up at their store and splurge with a good conscience.

Nevertheless, if not finding solutions to the information overload problem, this may prove be a major obstacle to the decommoditisation process. Simply because the consumers will not demand any additional information and just wish to treat goods as commodities.

10.2 Information Control: Decommoditisation Controlled!

A commoditisation assumption being made is that information will continue to travel freely, and that this trend will even be accelerated. Data proliferation is on the march, and there are even signs that freedom of data are becoming a central aspect in new semi-religious movement, where the general believe is that information freedom is a good in itself. However, even without religious zeal, large parts of the world has become accustomed to the free information flows which can be consumed, created and even altered by the general population.

However, not all governments or agencies agree whether complete freedom of information is such a great idea, and nations across the globe are already censoring many data streams. Some companies are also starting to do the same, both by direct order from governments but also on their own initiative to stem unwanted opinions and extreme movements. Some information curtailment may arguably be for the better, such as limiting distribution of extreme hate ideology or pedophile pornography.

However, the question is then always where the limit should be drawn, and who should be the gatekeepers. If internet would had been created in the early twentieth century, fascism and what we would brand today as extreme racist ideologies would have been common public discourse. It was not only the German nazis who was under the belief that white people were superior to others, but also vast parts of the American population, while the British Empire and French Republic arguably concurred. Large parts of the populations at this time also saw homosexuality as deviant and criminal, and even as late as the 1950s there were over 1000 men serving sentences in UK prisons for committing homosexual acts. 2000 years prior in ancient Mediterranean, non violent religious groups such as Christians where persecuted while pedophilia or "Pederasty" was considered socially acceptable under certain circumstances. So, internet censorship over the course of history would arguably had changed quite a lot over time, and the question is who should really be in charge of the censorship and whether these entities might be tempted to overstep their boundary under political pressure. If for instance some fringe environmental groups would become violent in a desperate attempt to save the planet, would not the regulating entity have a good case to pursue the environmental ideology that their crimes are based upon? If nationalistic economic policies are pursued by a government, would there not be a case for the regulating entity to limit access to overseas products and information, in favour of the domestic ones?

Furthermore, since the process of decommoditisation may entail some unflattering truths of certain countries industry practices, the fallout part from the economic effects could also have a political impact. There are thus plenty of reasons for government in conjunction with industry to curtail some of the information flows. This is already happening, with several countries doing what they can to manage information.

In China 2013, there was approximately 30,000–50,000 government employees involved in maintaining the great firewall, the common phrase for Chinese attempts to manage the internet flows of information. North Korea unsurprisingly have even more information control over the internet, but also nations like Iran and Saudi Arabia employs substantial internet censorship. Iran even rolled out a domestic internet in response to perceived western influence in 2015. Maybe more surprising is the extensive internet surveillance being conducted by nations such as the US and the UK. These countries do have relatively free internet access with limited censorship—but the extensive surveillance and limiting of information could easily be expanded in accordance with new political agendas in the future, since the infrastructure to do so is already in place.

Part from countries, companies such as Facebook, Youtube and Google already limits their content in response to public criticism, but the oversight on how this censoring is conducted is arguably very limited. A future scenario might be that these companies also starts to favour certain content from a business perspective, where large interest groups such as the incumbent industries gets more leverage than upcoming decommoditisers—which could put a cold blanket on the emerging trends.

Another possibility is that part from just concealing information; government and companies may also seek to falsify information to put them in a better light. This leads us to the next possible derailer—disinformation and loss of trust in information.

Knowledge Box: Our Information Diet

Many people put considerable effort and consciousness into what they eat, since this is known to have a significant impact on humans wellbeing. What people seem to care less about, but perhaps ought to care more about is what they also put into their brains in terms of information, as the kind of information we consume may have an outsized affect on our well being that we have yet to comprehend. Information quality have been known to be pivotal for kids learning and development, and many parents are very attentive of this

(continued)

fact by providing children with educative toys and TV shows, as well as maintaining a stimulating environment with good social context.

However, people seldom stops developing their mind just because they ceases to be recognised as kins. Grownup brains evolve over time as well, and much in accordance to what external stimulus it receives. This is rather obvious, as most people do not have the same opinions nor the same state of mind when they are 25 as when they are 45—but how the brain develops during those years in response to input is seldom something people pay a lot of attention to.

Consuming 1000 hours of science TV programs, will arguably have another impact on peoples mind than watching 1000 hours of celebrity shows, scary movies or the Jackass show. Not to say that either are superior to the other; 'Jackass' may bring a lot of joy to peoples lives, as could celebrity shows and scary movies. But over a time period of 20 years and thousands of hours of viewing, the mind will probably have evolved in quite a different direction whether the person focused their attention on the science program, scary movies or jackass—including the persons general approach to life.

A consumer can spend whole evenings watching 24 hour news media, which can serve the same story for hours without really communicating anything new of substance after the first few minutes. In addition to this, over half to the time watched may had been devoted to commercials. The person could in the same time period have tried to learn a new language, watched a great movie, knitted a pair of gloves or played guitar with their friends. All with arguably very different impact on the person's mind.

If junk food should be avoided, than maybe junk information should as well. People should also be mindful that many business models today are centred on keeping consumers 'hooked' and 'stealing' as much of the attention span as possible. Or as the TV company Netflix CEO once stated—our biggest competition is customers sleep! Maintaining a good information diet might thus be the single most pivotal health exercise a person can undertake—and be crucial to what person we will become in the future.

10.3 Consumer Disbelief: False Information and Greenwashing Poisoning the Information Flow

Information is a dish best served raw, since when cooked its becoming far more difficult to understand what the actual ingredients have been, and what's just industrial seasoning. Cooking information, putting an angle on a story or storylining is a very common practice across most markets, and everyone from NGOs and governments to companies and consumers own Facebook feeds are involved in the practice. In theory it is nothing wrong with this, as the information in itself served raw may not yield the necessary insight one wish to convey.

Insight can be supplied by putting data in a certain format together with other information without cherrypicking data, in order to show a certain trend, progression or to prove a point. This with the intent to help the receiver understand the implication of the information the person is looking at and to convey a certain message.

However, there is a subtle line between storylining information to provide insight, and storylining to angle information to suit ones own agenda. Its furthermore a very thin line between angling a story by excluding data that does not support your agenda, and simply make up information or supply information that is not based in solid science.

Common information hierarchy: Information → Knowledge → Wisdom → Bliss!

Marketing information hierarchy: Data → Insight → Story lining → Market share!

If excessive storylining, cherrypicking data and of limiting of information becomes widespread—distrust of information will follow. This has become an increasing problem with the onset of social media, where everyone is free to share any information including rumours with their peers. There is not anymore the filtering of information traditionally provided by media companies such as newspapers; who may be experienced to judge the validity of stories and ideally also accountable to make sure that the information they publish are true. Since peers are usually trusted, especially as they are usually holding similar viewpoint as oneself—the distrust of traditional media increases when conflicting with the viewpoints or the local information echo-chamber. Furthermore, modern media companies are usually profit driven—and their main goal will thus be to capture customer attention and supply the viewers with what they enjoy to see, hear and read. This unfortunately usually leans towards the spectacular, and most news channels can be sure to get more views of an MPs lewd sex scandal, than on an analysis of the long term economic effects from the population depletion of natural pollinators. Even though the latter will arguably have a more significant impact on the future life of the people watching, than the indiscretions of some arbitrary public figure.

The distrust in media reached a boiling point during 2016, in the run up to US presidential election, where the US electorate seemed completely unable to agree on what was fact and what was opinions. News outlets had became increasingly polarised and supplied conflicting views and data on the same topics, which made it quite difficult for anyone to understand what was actually going on, and how 'the other side' could think the way they did.

People may not want democracy, they might just want a dictator who agrees with them

Similar trends can be observed in companies as information suppliers, where many consumers are sceptical of the communication, since it is most likely created only to serve the company agenda. Green-washing has become a common phrase in this regard, where large polluters invest meager sums in fringe parts of the business, and communicate this vividly to make them look better. For example, a coal mining companies who puts up beehives on the roof of their main office to help pollinators, which may be good in its own right but may not offer a proper view of the business, if being the central focal point in the communicative effort as the 'sustainable bee company'.

Knowledge Box: The Data Religion

The data revolution, will have a dramatic impact on commodities in far reaching ways that we have still not yet grasped. There are even people believing that a so called data religion will take hold, where the main good is the unlimited flow of data—which will promote peace, happiness and the right choices of people if just dispersed enough. Proponents of this religion sees humans as conduits of data transfer, and increased knowledge and data uploaded to the internet through analysis will be what solves all ills. With just enough data and processing of that data, all problems can be solved. The holy grail for these datalogs is the invention of an algorithm that can understand humans better than humans can understand themselves. This algorithm, would then be used to govern humanity to produce results that are better for us than we could produce ourselves. This would include both economic output and democracy, which the algorithm would be much better suited to handle.

> If God did not exist, it would be necessary to invent him

If trust in information both from public and private sources becomes eroded, it will become increasingly difficult for companies to decommoditise their products. The consumers may not be willing to pay an additional cent for a mitigated externality, if they simply do not believe that the information they are receiving is accurate.

The distrust trend does not necessary have to impact the whole market though. Some states and companies will probably have more trust than others, and it might even be beneficial for some actors who are telling the truth if they are seen as one of the few in the market who does so. It takes a long time to gain trust, but only one mistake to deplete it. So if companies, nations or NGOs wish to decommoditise goods, they need to make sure that they gain credibility in the short run, but also that they keep that consumer trust over the long run in order to reap higher margins for their goods.

We have now looked at a few possible derailers of the decommoditisation trend. There may certainly be more than those described such as customer deprivation, negative communication fatigue, common greed and probably many others as well. That being said, most of the derailers do not seem to be out of the control for the decommoditisation stakeholders, but rather problems that can be managed, mitigated and eliminated if the right course of actions are taken. So even though decommoditisation should not be seen as an absolute certainty, there is definitely an opportunity to be cultivated on the horizon—which we will look further into in Chap. 11.

The Opportunities for a Better Tomorrow 11

We have in this book examined the history of commoditisation and the progression of decommoditisation so far, including what actors within the markets should do to reap the possible benefits. In this chapter will investigate what these general benefits might be.

11.1 The Opportunity of Decommoditisation

To be clear; companies should be able to make make more money out of the decommoditisation trend than they currently do selling undifferentiated bulk commodities, or the products containing them. The simple reason is that the decommoditised products should at least initially be more expensive to make and fetch higher prices than the bulk cousins. This entails increased revenues for companies involved in the value chain and if assuming same margins as for the bulk product, the bottom line of companies should simply increase. This may also be quite a bearish assumption since there should also be opportunities to actually increase margins on niche products compared to bulk commodities, due to the higher value added and less 'perfect competition' for the differentiated products on the market. The conceptual logic is visualised in Fig. 11.1.

In order to achieve higher margins or the same margins on top of increased revenue base, actors must invest in their brand capital. Strong brands which consumers trust will be key to fetch higher revenues on goods by infusing a sense of trustworthiness and thus increase value added compared to alternatives. However, this is not only true for companies, but also for countries. Made in Germany have for example become a very strong brand and arguably increased the value of most goods produced there—being usually associated with high quality, and in the future perhaps also with renewables due to the German renewable energy turn around (Energiwende).

Hence, it may become increasingly important for countries to clean up their act to brand themselves as a high value added producers, in order to enable their

F. Andrén-Sandberg, *Commodities Pricing and the Bulk Trap*,
https://doi.org/10.1007/978-3-319-72468-3_11

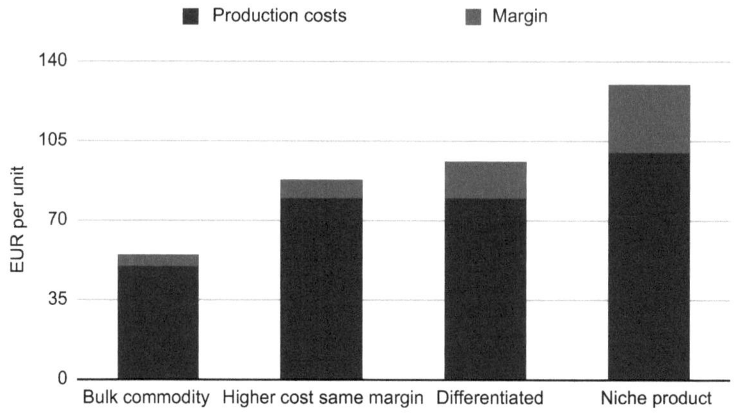

Fig. 11.1 Conceptual differentiated product logic. Market price for the example bulk commodity is EUR 55 with the margin of 10% on a EUR 50 production costs. If the same margin can be attained for a differentiated commodity with just higher overall costs—the 'Higher cost same margin' scenario with production costs EUR 80 would yield a price of EUR 88. A margin increase of 60% from EUR 5 → EUR 8. Assuming that a differentiated products could fetch a higher margin of 20%, or a niche product of 30%—the margin could grow by 320% to even 600%. The latter possibly an dream scenario, but serves to illustrates the logic that differentiated products should be able to increase margins for current actors involved in commodities

companies to better compete on the global market. I.e. countries with decent labour laws and good environmental policies might find a competitive edge in 'differentiated commodities markets'—and thus able to fetch higher margins than others because of their country of origin brand capital. This ideally raises income for the home country, who can then afford to progress standards even further. However, it might not only be good environmental behaviour that pays off, but also other traits and political standpoints which matters for some consumers. Mongolian horse milk, or Russian crafted iron may for instance inherit higher value for some customers compared to similar products made in Tuscany, simply because people do not usually associate Tuscany with either horses or iron but rather with nice food and wine.

Countries should in general also be aware that their political actions and those of its citizens can severely affect the prospects for other producers in that market. As an example; after France conducted nuclear tests in the mid 90s, there was a customer boycott against French wine in many countries, leaving bottles of wine on the shelves of foreign supermarkets and significantly hitting the French wine industry (British imports of French wines fell by over 30% within a year).

In Denmark 2005, a local newspaper published pictures of the profit Muhammed implying connections to terrorism and other negative stereotypes. This sparked protests and outrage across some countries in the Middle east, which caused embassies being torched and perhaps more strikingly; that many consumers started to boycott Danish products. This had a substantial effect on

Danish exports, as dairy products had become an increasingly important revenue source for Denmark from the middle east.

In the USA 2003, in the heated atmosphere after the 2001 New York terror attack, some American consumers stopped buying French products since France did not support the Iraq war. French fries was even relabelled 'Freedom fries' in some restaurants (including us US congress cafeterias), even though it did not affect France in any material sense. As a curiosity, similar things happened during WW1 when the German Sauerkraut was renamed liberty cabbage in America. Even the sausage slang word 'hot dog' got an upswing in the same period as it was previously commonly named Frankfurter—hot dog has arguably been going strong ever since.

Nevertheless, it is not to say that the countries in the former examples did anything wrong, nor that they should refrain from taking the political decisions that they feel is right. But it just shows that the political behaviour of countries matters for consumers, and might do even more so in the future when people get access to increasingly transparent information streams regarding the origin of their potential purchases. This may be a very positive development since countries with poor human rights records, insufficient labour legislation, aggressiveness towards neighbours or other unethical policies might be punished by conscious consumers—and actually provide an incentive barrier towards negative behaviour. It also highlights the importance of country branding, which may provide a significant competitive advantage in the future. The trend does entail a risk that countries with strong media resources can mislead consumers and distort available information to promote its own causes. However, this would not be much material different than the current state of affairs, and increased information transparency together with dispersion of information sources should make this practice increasingly difficult.

11.2 Why Should We Embrace Decommoditisation?

Consumers tend to vote with their feet's towards countries, certificate bodies and brands. The main task for companies regardless of where they are situated in the value chain, must thus be to create goods that supports their customer's current pallets and to take advantage of the opportunity that the latest trends presents.

In regards to the decommoditisation trend, this implies the need for constantly researching the market and analyse the fit with the current product portfolio, put the data into practice and work across the value chain with other actors to support business development. This may require strong collaboration between companies, NGOs, governments and other stakeholders, and might even usher in a new economic paradigms of integrated production and after sales services of products. Certification schemes could be created where they are not yet present, and new integrated information platforms for the products formed.

This decommoditisation trend will have both winners and losers, and market forces can be brutally efficient towards those companies and countries which do not adapt to new market conditions. Evolution does not ensure the survival of the strongest, but rather the one most adaptive to change.

The trend might also be what pivots modern societies towards the so called 'circular economy', where products and commodities ceases to be used in a cradle to grave manner, and are instead recycled into new products at the end of its lifetime. This change may not just be a new 'feel good' production mode, but rather a necessity in order for global economic growth to continue.

Commodity consumption has increased dramatically over the last century, and with another 2 billion people awaiting to join the middle class within the next two decades, there may simply not be enough raw materials out there to sustain such an increase without changing the way primary resources are utilised. The main viable solution to continue our global standard of living increase is to recycle the materials that we use—where old plastics products can become new plastic products and food waste nutrients is being collected and recycled at our farms to produce new crops and food.

The simple fact coming out of the statistics is that even if there would not be another 2 billion people joining the middle class, we would still be running out of easily accessible resources at current consumption rates, and the linear business model will arguably soon reach its limited timespan in history. An interesting note is that most economies historically used to be circular economies, and the linear model is only a very recent phenomena. However, the current linear model might soon prove to be nothing more than a short parentheses in history. As although being a period of great human progress, also a very inefficient and wasteful epoch only possible due to insufficient information flows, short term thinking and lacking knowledge among policy makers, companies and consumers alike.

That being said, nothing is set in stone. Since the main driver of decommoditisation is actually not technology such of the internet and social media, but rather a culture who cares and refuses to accept unethical or unsustainable production practices, or for that matter where a product is produced. This could of course change, where people might instead start to feel climate and sustainability fatigue as an effect of too much negative information—causing people to simply stop caring about the issues.

It might also become the case as explored in Chap. 10, that strong economic forces not wanting to see this trend, mobilises everything in their arsenal to derail it, and succeeds. Maybe censorship of information and concealment of facts will work to obscure reality from consumers—where powerful states who do not want their stakeholders to see the human conditions in their production facilities, starts to limit and punish information spreading in the name of national security.

These negative scenarios are all possible, but it will be difficult for even the most powerful of governments to fight such strong technological paradigms as the internet and social media is ushering forward. Furthermore, if one would look for the historical odds of doing so, the momentum is clearly on the side of justice and technological progress—where heavy censorship and banning of technologies have rarely budged the long term trend of human progress.

11.3 Why This Is a Good Thing

If the decommoditisation trend gathers pace, consumers will be able to purchase according to their knowledge and values, and if their values are good the world will probably become a better place.

From a wider perspective, decommoditisation could even be seen as a further step in the democratisation process, where determination of production modes can be managed by all who consume the products and services. It will at least enhance the possibilities for people in non-democratic but market oriented countries, to gain more leverage over their nations development by channeling their demand towards the production modes that they wish to see in their own country.

Disregarding far-fetched hopes, decommoditisation might at least lead to a more conscious sourcing of current commodities, as well as increase recirculation of inputs which may become a vital part of every producers sourcing portfolio.

The need for the global economy to change into more sustainable souring of commodities should thus not be seen as a matter of empathic considerations, but rather out of pure necessity—since the current path of sourcing and treat commodities entails that our economies will simply run out.

It is not that humanity need to stop using commodities, but our economies do need to make sure that they are replenished at the same pace we use them, and ensure that the extraction does not cause more negative than positive overall economic impact over the long run. Its not difficult logic to understand that ten trees cannot be cut down per day over the long run if only five are replanted—and that there's a need to manage resources well to ensure long term prosperity of our civilisation.

The good news is that through decommoditisation, we've actually been given one of the levers to change the somewhat negative resource downward spiral. Imagine that this generation may become the generation that actually reverses global warming, the generation who leaves more fish in the ocean and more trees in the forest compared to when they were born, and to become the first generation in arguably more than 300 years who actually leaves the planet in a better resource state to our children, than how we inherited if from our parents.

This is something to be really excited about, and with business models and technology supporting the trend—this really is the opportunity for a better tomorrow!

12 Epilogue and Further Reading

Decommoditisation come as it may, is a trend which we can see currently in the market—but all forecasts are knowingly very difficult to conduct, especially those which concerns the future. There are strong evidence that this trend will continue, but history show's that there is always room for the 'Deus Ex Machina' or 'Diabolus Ex Machina' depending on the impeding outcome and your given interest.

What will happen is uncertain, and the best we can do today is of course to brace us for that decommoditisation future. Or perhaps not! We are in facts ourselves not only passengers on this journey, but we are also the creators of the future. If humanity wish to have decommoditisation, then decommoditisation will be become the future. If humanity decides that decommoditisation is nothing to be reckoned with, then it will certainly not become the future. The most likely scenario is a trade off in-between the two, but it will in the end be up to choices made by consumers, companies and policymakers to what level they wish to see this trend to proceed.

There is furthermore plenty of room for individuals, and individual companies to have an outsized impact on this progression. Historic trends is seldom decided by the masses, but usually by a few people or organisations who influences the masses into a whole new direction.

Humans are the only animal known which can cooperate flexibly with other individuals of the flock in the order of thousands. Ants can also cooperate in the thousands, but not flexibly as they are stuck into a very specific behavioural pattern. Some apes can also cooperate, but only in numbers maximising approx 150–250 individuals, due to the maximum social connections an individual ape can maintain. What makes humans special is that we instead of direct social connection, can cooperate through abstract ideas such as religion, companies or even an idea such as decommoditisation. And if just enough people decides to vote with their feet, influence their friends, envision new thoughts and even starts up companies in pursuit of those thoughts—it can have an enormous impact on the future of the global economy.

F. Andrén-Sandberg, *Commodities Pricing and the Bulk Trap*,
https://doi.org/10.1007/978-3-319-72468-3_12

It seems like many people have already decided to move along the decommoditisation line of thinking, and the rest of us better catch up if wanting to influence how the next economic evolution is going to shape our common reality.

If you are further interested in commodities or decommoditisation, there are several articles in the area that can be accessed through any good publishing database. There are also plenty of good books regarding commodity trading, which gives a good overview of the current 'fundamentals' and beliefs of the commodities market. There also needs to be further research conducted within this field, and there will be plenty of opportunities present for anyone who wish to investigate the trend more in depth—and maybe even falsify some of the current believes which will lead us to new insights into the processes.

The exploration of decommoditisation is evolving and any input could be of high value to the current market understanding. If you are interested to join the discussion and post your own inputs to this emerging field you are welcome to join the discussion at: EndOfCommodities.com or log onto the Facebook group EndOfCommodities. Where you will also be able to find the latest updates and information within the field.

Thank for reading and best wishes on your decommoditisation journey!

MIX
Papier aus verantwortungsvollen Quellen
Paper from responsible sources
FSC® C105338

If you have any concerns about our products,
you can contact us on
ProductSafety@springernature.com

In case Publisher is established outside the EU,
the EU authorized representative is:
Springer Nature Customer Service Center GmbH
Europaplatz 3, 69115 Heidelberg, Germany

Printed by Libri Plureos GmbH
in Hamburg, Germany